HVAC COMMISSIONING GUIDEBOOK

Green buildings have become common in India and other countries in Asia. However, there is a concern regarding the performance of green buildings failing to meet the expectations of clients during the operation. One of the key reasons for this is poorly commissioned HVAC systems.

In this publication we provide tools and knowhow for more efficient HVAC commissioning. It gives answers for four major questions: why commissioning is needed, how to perform proper commissioning, which key performance issues of common HVAC equipment need to be considered, and what kind of checklists are used during commissioning? It covers the entire commissioning process beginning with the owner's project requirements and commissioning design reviews. Then, it explains procedures during installation and start-up of equipment followed by the functional performance testing, seasonal commissioning and 10 months' operation review.

This publication is developed by Indian Society of Heating, Refrigeration and Air Conditioning Engineers ISHRAE for Indian and Asian requirements in conjunction with the Federation of European HVAC Associations REHVA. The process steps described in this publication are in line with all major international building standards and green building certification schemes.

ABOUT THE AUTHOR

Maija Virta (M.Sc.Eng) is the Founder Director of the Santrupti engineers Pvt Ltd. She has over 30 years of experience in construction and HVAC-industry around the world and she has been in India for the last 7 years. Maija's key areas of expertise are indoor air quality, building commissioning as well as sustainable building policies and technology. Prior to moving to India, Maija was the CEO of the Green Building Council Finland. She has also been the vice-president of REHVA and the president of Finnish HVAC association SULVI. Among her various contribution to this field, she has authored many books and publications on sustainable buildings and technology - including the ISHRAE-REHVA HVAC Commissioning Guidebook. She has lectured on innumerable topics for ISHRAE, REHVA and ASHRAE. She is the member of two ISHRAE task forces: the IEQ task force that has prepared the IEQ standard for India and the HVAC Commissioning task force that is preparing the ISHRAE commissioning standard.

HVAC Commissioning Guidebook

Maija Virta, M.Sc Eng.,
The Founder Director
of the Santrupti engineers Pvt Ltd.,

CRC Press
Taylor & Francis Group
Boca Raton London New York

CRC Press is an imprint of the
Taylor & Francis Group, an **informa** business

REHVA

ISHRAE®

Federation of
European Heating,
Ventilation and
Air-conditioning
Associations

First published 2021
by CRC Press
2 Park Square, Milton Park, Abingdon, Oxon, OX14 4RN

and by CRC Press
6000 Broken Sound Parkway NW, Suite 300, Boca Raton, FL 33487-2742

© 2021 ISHRAE

CRC Press is an imprint of Informa UK Limited

The right of Maija Virta to be identified as author of this work has been asserted by him in accordance with sections 77 and 78 of the Copyright, Designs and Patents Act 1988.

Print edition not for sale in South Asia (India, Sri Lanka, Nepal, Bangladesh, Pakistan or Bhutan).

British Library Cataloguing-in-Publication Data
A catalogue record for this book is available from the British Library

Library of Congress Cataloging-in-Publication Data
A catalog record has been requested

ISBN: 978-0-367-75707-6 (hbk)
ISBN: 978-1-003-17301-4 (ebk)

ISHRAE®

FOREWORD

This HVAC Commissioning guidebook is a very important and much-awaited publication that can help all stakeholders interested in delivering quality buildings that meet the owner's requirements from Day one.

It is the first guidance on this level for us who have our daily work with Building Commissioning.

Maija Virta has collected material related to The Commissioning Process, as well as the technical aspects of Commissioning of various systems, and entered it in a framework as described by the International Energy Agency in ECBCS Annex 47. The result is a guidebook that describes a process compatible with the routines in the building sector almost everywhere on our globe. This is the first work that in this way describes both the process in a very hands-on manner and details the commissioning activities for various types of systems, complete with theoretical background, guidance & checklists.

I don't think any other person than Maija could have done this. Maija has a long history with REHVA as author of several REHVA Guidebooks, author of numerous articles on indoor climate and specific installation types, as a scientist and as a hands-on practitioner when it comes to troubleshooting sick-buildings and Building Commissioning. This rare combination of intellectual and practical skills is adding the rare quality of direct applicability on almost any Building project for the skilled Commissioning provider.

Maija is now living and working in India, but the knowledge accumulated in this guidebook is universal. There can be minor differences between what is customary in the various countries when it comes to the deliveries of the designers and contractors, and it can have influence on the wording in contracts in the different countries. However, going through the contracts are necessary anyway to make these documents aligned with the individual building project.

The Guidebook describes relations between Commissioning and the sustainability certification program LEED from US Green Building Council. It makes sense because LEED is widely used in India, but the described techniques can be applied to projects working with other sustainability programs, for example BREEAM, DGNB or HQE.

This is a HVAC guidebook, but the process described can be applied to the building enclosure as well. To complement the process described here, the checklists and other practical guidance from NIBS Guideline 3-2012, Building Enclosure Commissioning Process, no. 52 in the references, can be applied.

Both in India and in Europe we build far too many dysfunctional buildings. With this publication there no more is any excuse; - take a deep breath, and get started with Building Commissioning!

Ole Teisen

REHVA Commissioning Task Force Chair

Chief Consultant, Sweco Danmark A/S

PREFACE

Efficient building performance is more important than ever due to the changes in people's expectations and the business environment.

People have become aware of issues impacting their health and wellbeing. Built environment plays a key role in protecting our health from continuously worsening outdoor conditions like excess heat, air pollution, rain and noise. It is also important that buildings do not become the source of pollution and health hazards. At the same time, the amount of resources like water and energy are decreasing and the cost of their usage is increasing.

Investors are expecting higher returns on their investments. The latest building valuation and management models are linking the building performance as a key element in arriving at the evaluations like the discounted cash flow method calculating building value, life cycle assessments (LCA) and costing (LCC) as well as the holistic use of Building Information Modelling (BIM). As better performance leads to the higher value of a building, investor is ready to invest more money for achieving better performance results. This creates more pressure in the industry to improve the commissioning and building evaluation practices.

Technology used in the buildings have become more complicated and automated. This is why the start-up of both the individual components as well as the total system operation is crucial to deliver the benefits promised by manufacturers to the client.

Today all major green building certification systems have the HVAC system commissioning as a prerequisite. However, proper system performance should be the basic requirement in every new construction and refurbishment project.

The journey of creating this guidebook has been one of enlightenment for me as I have learned immensely about the efficient building performance and commissioning process steps both from my Indian and European colleagues. I would like to acknowledge the inputs that I received from the industry, especially from the following people: Mr. Nirmal Ram and Mr. Pankaj Shah for the vision in ISHRAE of promoting the HVAC commissioning, Mr. Suhaas Mathur for his technical contribution when the content was created and ISHRAE team for reviewing and giving valuable feedback - especially Mr. Nitin Deodhar, Mr. Sunil C Karandikar and Mr. Jitendra M Bhambure along with his team. I would also like to recognise the REHVA team i.e. Ole Teisen, Zoltan Magyar, Eric Melquiond and Frank Hovorka who has contributed in terms of sharing the vision and inputs regarding the commissioning process.

I hope that this guidebook will be a valuable tool to better building performance thus improving our health and endeavours.

New Delhi, August 2017

Maija Virta

LIST OF CONTENTS

ABBREVIATIONS

AC = Air Conditioning

AHU = Air Handling Unit

ANSI = American National Standards Institute

ASHRAE = American Society of Heating, Refrigerating and Air-Conditioning Engineers

BACS = Building Automation Control and Systems

BAS = Building Automation System

BECx = Building Enclosure Commissioning

BEE = Bureau of Energy Efficiency, Ministry of Power, Government of India

BIM = Building Information Modelling

BIS = Bureau of Indian Standards

BMS = Building Management System

BoD = Basis of Design

BTU = British Thermal Unit

CEN = European Committee for Standardization

CFC = Chloro-Fluoro-Carbon refrigerant

CFD = Computational Fluid Dynamics

CH_2O = Formaldehyde

CHW = Chilled Water

CHWST = Chilled water set point temperature

CO = Carbon Monoxide

CO_2 = Carbon Dioxide

CFR = Current Facility Requirement

CTC = Cost to Company salary

Cx = Commissioning

CxFR = Commissioning Final Report

CxP = Commissioning Process

CxPR = Commissioning Progress Report

DB = Dry Bulb temperature

DDC = Direct Digital Control

DHW = Domestic Hot Water

DLP = Defects Liability Period

DOAS = Dedicated Outdoor Air System

DOE = Department of Energy in the United States

DP = Differential Pressure

D = Diameter

DX = Direct Expansion cooling coil

ECBC = Energy Conservation Building Code, India

ECO = Energy Conservation Opportunities

EXWT = Entering Condenser Water Temperature

EML = Equivalent Melanopic Lux

EMS = Energy Management System

EN = European standard published by CEN

EPI = Energy Performance Index

$ePM_{1/2.5/10}$ = Particulate filtration classes as per ISO 16890 standard

FACP = Fire Alarm Control Panel

FAS = Fire Alarm System

FPT = Functional Performance Test

GRIHA = Green building rating scheme by TERI, India

H_6C_6 = Benzene

HC = Hydro-Carbon refrigerant

HCFC = Hydro-Chloro-Fluoro-Carbon refrigerant

HFC = Hydro-Fluoro-Carbon refrigerant

HCHO = Formaldehyde

HVAC = Heating, Ventilation and Air Conditioning

IAQ = Indoor Air Quality

IEQ = Indoor Environmental Quality

IGBC = Indian Green Building Council

ISO = International Organization for Standardization

IRC = In-row cooler in data centre

IST = Integrated System Testing

ISHRAE = Indian Society for Heating, Refrigeration and Air Conditioning Engineers

KMnO4 = Potassium permanganate

KPI = Key Performance Indicator

LCC = Life Cycle Cost

LCA = Life Cycle Analysis

LCHWT = leaving chilled water temperature

LEED = Green building rating scheme by USGBC

L_{eq} = Equivalent Continuous Sound

LRA = Locked Rotor Amps

MCC = Motor Control Centre

MEP = Mechanical, Electrical and Plumbing

MERV = Minimum efficiency reporting value

M&V = Measurement and Verification

NABERS = Australian green building assessment tool

NADCA = National Air Duct Cleaners Association, US

NBEE = National Environmental Balancing Bureau, US

NC = NC noise criteria

NH_3 = Ammonia

NO_2 = Nitrogen Dioxide

NR = NR noise criteria

O_2 = Oxygen

O_3 = Ozone

OCx = Ongoing Commissioning

ODP = Ozone Depletion Potential

O&M = Operation and Maintenance

OPR = Owner's Project Requirements

PAC = Precision Air Conditioning

PCO = Photocatalytic Oxidation

PEX = Cross Linked Polyethene plastic pipe

PE-RT = Polyethylene of Raised Temperature Resistance plastic pipe

PHI = Photo-hydro-ionization

PM = Particulate Matter

PID = Proportional–Integral–Derivative controller

R11, R12,... = Different types of refrigerants

RAMA = Refrigeration and Air Conditioning Manufacturers Association of India

RCx = Re-Commissioning

REHVA = Federation of European Heating, Ventilation and Air Conditioning Engineers

RH = Relative Humidity

RLA = Rated Load Amps

ROI = Return on Investment

RSPM = Respirable Suspended Particulate Matter

SO_2 = Sulphur Dioxide

SOO = Sequence of Operations

SOP = Standard Operating Procedure

SPLA = A-weighted Sound Pressure Level

SPM = Suspended Particulate Matter

STI = Speech Transmission Index

TBC = Total Bacterial Count

TBCx = Total Building Commissioning

TEWI = Total Equivalent Warming Impact

TFA = Treated Fresh Air unit

TFC = Total Fungal Count

THR = Total Heat of Rejection

TVOC = Total Volatile Organic Compound

UGR = United Glare Rating

USGBC = Green Building Council in the United States of America

UVGI = Ultraviolet Germicidal Irradiation

VAV = Variable Air Volume

VCD = Volume Control Damper

VFD = Variable Frequency Driver

VOC = Volatile Organic Compound

VRF = Variable Refrigerant Flow

VRV = Variable Refrigerant Volume

WB = Wet Bulb temperature

WELL = Green Building rating standard by International WELL Building Institute (IWBI)

WHO = World Health Organisation

TERMINOLOGY[14]

Balancing

The process of adjusting the flow rates of a fluid (air or water) in a distribution system to achieve the design flow rates within the specified tolerances.

Basis of Design (BOD)

A document that records the concepts, calculations, decisions, and product selections used to meet the Owner's Project Requirements and to satisfy applicable regulatory requirements, standards, and guidelines. The document includes both narrative descriptions and lists of individual items that support the design process.

Building Enclosure Commissioning (BECx)

It is a holistic process to ensure that exterior enclosure of building meets the Owner's Project Requirements regarding the material compatibility, continuity of structural layers, thermal and moisture performance, air tightness, etc.

Checklist

Project and element-specific checklists that are developed and used during all phases of the Commissioning Process to verify that the Owner's Project Requirements are being achieved. Checklists are used for general evaluation, testing, training, and other design and construction requirements.

Commissioning (Cx)

Project and component-specific checklists that are developed and used during the Commissioning Process to verify that the Owner's Project Requirements are being met. The checklists are used for general evaluation, testing, training, and other design and construction requirements. Commissioning (Cx) Commissioning describes a systematic process from the planning stage up to a time of about one year after completion of the building, ensuring by means of testing and documentation that all building services installations, both individually and interconnected, function in accordance with the Owners Project Requirements, Basis of Design and planning documentation. Commissioning extends up to a point in time during actual building operation so as to allow optimisations of operational parameters under actual load conditions and a range of climatic influences.

Commissioning Process (CxP)

A quality-focused process for enhancing the delivery of a project including Commissioning Process activities specified by the owner or required by a code or standard. The process focuses on verifying and documenting that all of the systems and assemblies under the scope of commissioning are planned, designed, installed, tested, operated, and maintained to meet the Owner's Project Requirements.

Commissioning Final Report (CxFR)

A document that records the activities and results of the Commissioning Process and is developed from the final Commissioning Plan with all of its attached appendices.

Commissioning Plan (CxPlan)

A dynamic document that outlines the organization, schedule of activities, allocation of resources, and documentation requirements of the Commissioning Process. The document shall be updated frequently, at least at the end of every phase, with the tasks and documents related to the coming phase.

Commissioning Progress Report (CxPR)

A written document that describes the activities completed as part of the Commissioning Process and significant findings from those activities. The commissioning progress report is continuously updated during the course of a project.

Commissioning Provider (or Authority (CxA) or Agent)

An organization or individual identified by the owner who sets up the Commissioning Team and leads, plans, schedules, and coordinates the activities to implement the Commissioning Process.

Commissioning Team (Cx Team)

The individuals and agencies who, through coordinated actions, are responsible for implementing the Commissioning Process.

Commissioning Testing

The evaluation and documentation of the delivery condition, installation, and proper functioning of the equipment and assemblies according to the manufacturer's specifications, and project documentation to meet the criteria in the Owner's Project Requirements.

Design Review—Commissioning

A review of the design documents to determine compliance with the Owner's Project Requirements, including coordination between systems and assemblies being commissioned, features and access for testing, commissioning and maintenance, and other reviews required by the OPR and Cx Plan.

Design Review—Peer Review

An independent and objective technical review of the design of the project or a part thereof, conducted at specified stages of design completion by one or more qualified professionals, for the purpose of enhancing the quality of the design. (Not part of Cx Process)

Functional Performance Testing (FPT) (or Integrated System Testing (IST))

Functional testing of the interactions between systems that need to function together to meet the OPR and BOD. All systems must run in the automated mode as described in the design. FPT plan defines methods, steps, personnel, and acceptance criteria for tests conducted on different systems and interfaces among system.

Installation Review

Observations or inspections that confirm the system or component has been installed in accordance with the contract documents and to industry accepted best practices.

Commissioning Log-book

An official and on-going record of issues or concerns occurring during the course of the Commissioning Process and their resolutions that have been compiled by members of the Commissioning Team.

Ongoing Commissioning Process (OCx)

A continuation of the Initial Commissioning Process into occupancy phase to continually improve the operation and performance of a facility to meet current and evolving Owner's Project Requirements or Current Facility Requirement (CFR). On-going Commissioning Process activities occur throughout the life of the facility.

Owner's Project Requirements (OPR)

A written document that details the functional requirements of a project i.e. the expectations of how it will be used and operated. It also includes project goals, measurable performance criteria (for e.g. energy efficiency goals), cost considerations, success criteria and relevant benchmarks, and supporting information.

Re-Commissioning (RCx)

An application of the Commissioning Process requirements to a project that has been delivered using the Commissioning Process.

Retro-Commissioning

The Commissioning Process applied to an existing facility that was not previously commissioned.

Start-Up Tests

Tests which validate that a component or sub-system is ready for automatic operation in accordance with the manufacturer's requirements.

System O & M Manual

A system-focused composite document that includes the design and construction documentation, facility guide and operation manual, information on maintenance and facility personnel training, Commissioning Process records, and additional information of use to the owner during occupancy/operations.

Test Procedure

A written protocol that specifies the methods, personnel requirements, and expected outcomes for tests conducted on components, equipment, assemblies, systems, and interfaces among systems to verify compliance with the Owner's Project Requirements.

Total Building Commissioning

A systematic process of assuring by documentation and verification that all facility systems perform interactively in accordance with the design documentation and intent, and in accordance with the owner's operational needs, including training of operation personnel. The process starts from the design phase and continues up to a minimum of one year after construction.

1. COMMISSIONING IN A NUTSHELL

1.1. What commissioning is and what it is not?

Commissioning is the process of assuring that all systems and components of a building are designed, installed, tested, operated, and maintained according to the operational requirements of the developer, owner or end user.

In practice, the commissioning process consists of procedures to check, inspect and test every operational component of the project, both their individual functions and their integration as sub-systems and performance as complete system. Commissioning is one of the core activities (Fig.1.2.) to ensure good indoor environment for users, sustainable building operation, increased value of investment and reduced environmental load.

It guarantees proper system performance and renders the desired benefits of an efficient system design. Commissioning is a systematic process of ensuring that all building systems perform interactively according to the design intent and the owner's operational needs. Initial commissioning is a systematic quality assurance process which through planning, testing and verification, ensures that building meets the Owner's Project Requirements (OPR) and the owner achieves a successful construction project.

Seasonal commissioning is needed to fine-tune system operation in different seasons and ensure that all operational failures are found and corrected during the first year of operation. Ongoing commissioning helps to continuously monitor operation over a life time and plan corrective actions both for short and long term. (Fig. 1.1.) The economy of operation and user comfort are improved as the building operates in different seasons with optimum operational parameters.

Proper commissioning also contributes to the user's health. In areas where ambient air quality is poor, properly performing systems keep the unwanted pollution outside the building. As the commissioning process also focuses on training of operation and maintenance (O&M) personnel, it is more likely that systems are operated in such a way that a safe and healthy environment for users is achieved by, e.g. avoiding legionella or mould problems.

The commissioning process ensures that knowledge and resources are available to maintain systems as they were originally intended and reduce the number of problems upon initial occupancy.

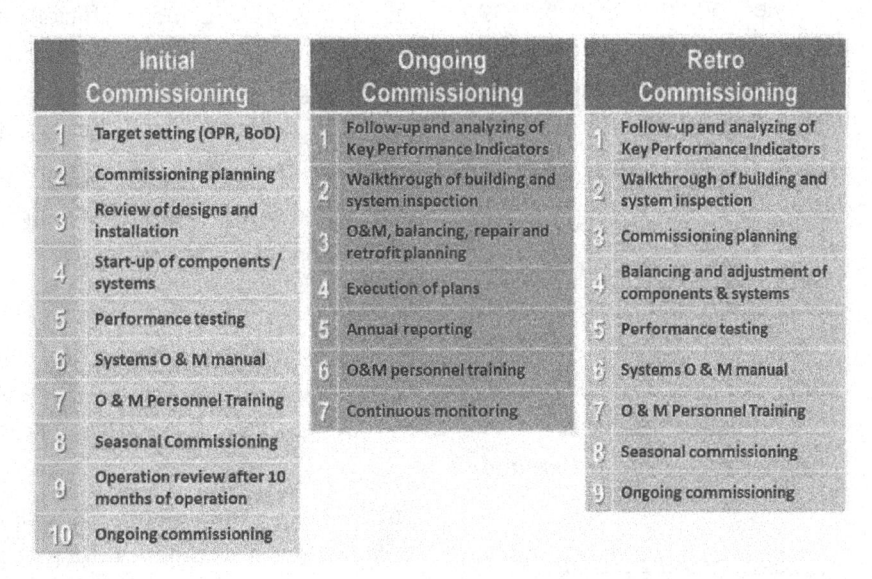

Initial Commissioning		Ongoing Commissioning		Retro Commissioning	
1	Target setting (OPR, BoD)	1	Follow-up and analyzing of Key Performance Indicators	1	Follow-up and analyzing of Key Performance Indicators
2	Commissioning planning	2	Walkthrough of building and system inspection	2	Walkthrough of building and system inspection
3	Review of designs and installation	3	O&M, balancing, repair and retrofit planning	3	Commissioning planning
4	Start-up of components / systems	4	Execution of plans	4	Balancing and adjustment of components & systems
5	Performance testing	5	Annual reporting	5	Performance testing
6	Systems O & M manual	6	O&M personnel training	6	Systems O & M manual
7	O & M Personnel Training	7	Continuous monitoring	7	O & M Personnel Training
8	Seasonal Commissioning			8	Seasonal commissioning
9	Operation review after 10 months of operation			9	Ongoing commissioning
10	Ongoing commissioning				

Figure 1.1. Commissioning starts as initial commissioning during the construction project or as retro-commissioning in an existing building that has not been commissioned before; seasonal commissioning is carried out during the first year of operation and continuous commissioning is carried out annually over a building's lifetime.

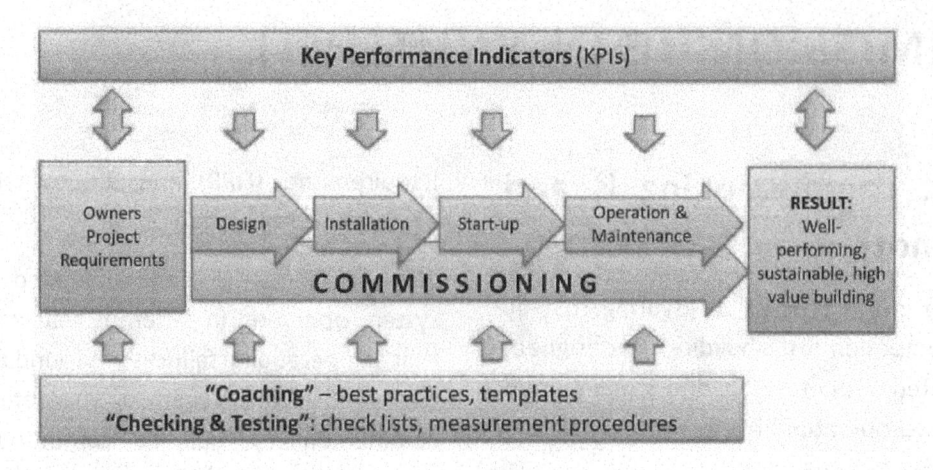

Figure 1.2. Commissioning Process is a support process to assist the main processes to design, construct and operate a building.

The Commissioning Provider conducts quality checks of design, installation and start up procedures and ascertains that proper operation is achieved. It is essential to run the functional performance tests, where the entire building and system operation is evaluated under different conditions, besides ensuring that operation and maintenance personnel are ready to operate the building.

A commissioning process focuses first on system operation and second on the kind of equipment required to achieve the operation. It documents all problems and errors that contradict the OPR and devises their solutions in a structured way. Ideally, commissioning provider should be an independent company that has no affiliation with designers, contractors, manufacturers of equipment or system components or any other person or company which could affect the ability of the Commissioning Provider to generate an independent commissioning report.

Commissioning is not a replacement for the existing quality inspection process during design and construction but it is an addition to these processes. It is also not an additional phase to the existing pre-design, design, construct, occupancy and operation phases; it runs parallel with the core processes. Commissioning is not an isolated testing event of a single equipment, neither is it an adjusting and balancing tool.

Rather, it is the total process focusing on the entire system operation such that all subsystems perform well together and optimum building operation is achieved.

1.2. Who needs commissioning?

Proper commissioning brings benefits to all stakeholders: users, owners, developers, investors, designers, contractors and manufacturers. It is important to understand and communicate the value addition to all stakeholders. (Fig. 1.3.)

The occupants become more productive as thermal comfort and indoor air quality get better. Therefore, a tenant is willing to pay higher rent to the owner. As owners have a positive cash flow via reduced operating costs, increased rent and an increased building value, their ability to invest in proper commissioning and well-performing systems increases and they will be inclined to employ on independent commissioning authority.

Occupants will be more efficient, productive and healthier in buildings that perform properly and where O&M teams are aware of the connection between system operation and occupants' comfort. Owners and developers get what they have specified in the OPR, whether it is high energy efficiency, good indoor conditions or low operational costs.

Investors also benefit as a well performing building is value for money, either through lower operational costs, higher rent or increased value of the building. Design team (e.g. architect and MEP designers) can satisfy the owner with a well performing building. Efficient commissioning process also ensures less changes and delays during the taking over of systems. A satisfied building owner is more likely to give work in the future. A contractor can deliver and demonstrate that what was promised has been delivered. There is also less hassle and delays during handover and the Defects Liability Period (DLP).

Component manufacturers are constantly developing new and sophisticated products to achieve increasingly demanding system performance targets set by clients. With proper commissioning they can demonstrate the benefits of their products when the project becomes operational.

1.3. 'Mandatory' commissioning in green buildings and beyond

Commissioning (Cx) activities are not mandatory by building regulation in most of the countries. However applicants of Green Building Certifications like LEED or IGBC Green have to demonstrate certain minimum commissioning activities.

Some Green Building Certification (e.g. LEED and IGB Green) requires fundamental commissioning as a prerequisite to support the design, construction, and eventual operation of the project to meet the Owner's Project Requirements (OPR) for energy, water, indoor environmental quality and durability.

The owner has to document the requirements as an Owner's Project Requirements (OPR) document and designers have to develop the Basis of Design (BoD) document. The owner may also nominate the independent Commissioning Provider.

The Commissioning Provider must complete the following commissioning activities for mechanical electrical, plumbing and renewable energy systems and assemblies as follows:

- Review OPR and BoDs;
- Develop and implement Cx plans, checklists & functional performance test procedures;
- Confirm incorporation of Cx requirements into the construction documents;
- Verify the system performance.

Document all findings and recommendations and report them directly to the owner & maintain logbook throughout the process. There are numerous commissioning activities that are beneficial in terms of building operation and meeting the owner's project requirements. Many of these are also required to get additional green building points in certification systems.

For example, LEED Enhanced Commissioning credit requires review of contractor submittals, creation and verification of System O&M manual, inclusion of operator and occupant training, seasonal commissioning and review of system operation after 10 months of operation. Commissioning is required throughout the lifetime of the building. Therefore planning of on-going commissioning process during the initial commissioning process is important.

A building's thermal envelope also requires commissioning as its enclosure consists of various materials and involves several stakeholders. The Building Enclosure Commissioning (BECx) is utilized to validate that the performance of materials, assemblies, components, systems and design fulfils the owner's requirements.

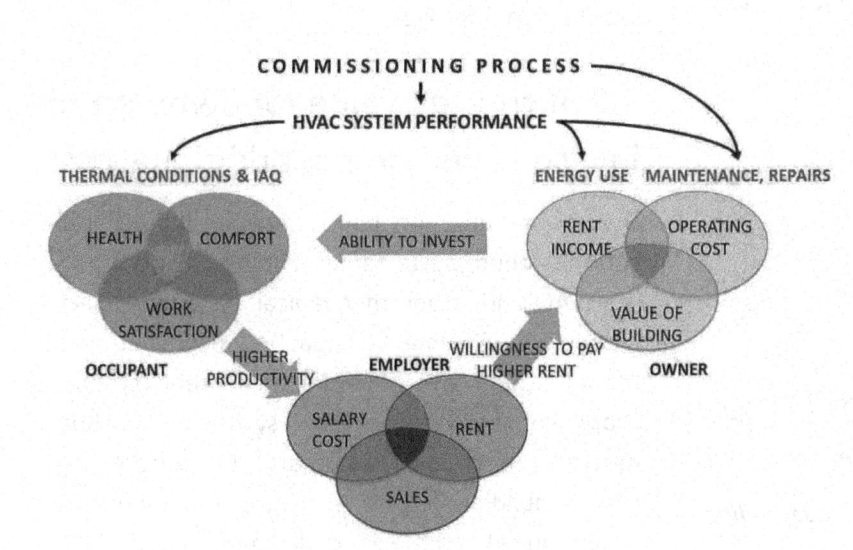

Figure 1.3. Well performing (commissioned) building is Value for Money to all stakeholders.

2. WHY COMMISSIONING?

2.1. Proper performance ensures healthier and more productive occupants

HVAC and other mechanical systems play a key role in creating a healthy and comfortable work environment. Building commissioning focuses on proper system validation and operation. Therefore, it ensures that thermal conditions, indoor air quality, lighting and acoustic conditions are met as specified in Owner's Project Requirements. Discomfort increases the stress hormone level and leads to various illnesses. In a comfortable environment, not only are people healthier but also function more effectively.

The World Health Organization (WHO) report titled 'The Right to Healthy Indoor Air' (2000) states the following: 'Indoor air quality is an important determinant of population health and wellbeing. Exposure to the hazardous airborne agents present in many indoor spaces causes adverse health effects such as respiratory diseases, allergy and irritation of the respiratory tract. That is why our responsibility is to provide healthy indoor environment for occupants. [75]

In an office building, the salary paid to the staff is the biggest cost over the building's lifetime. There is scientific proof that good indoor environmental conditions improve a worker's health and productivity.

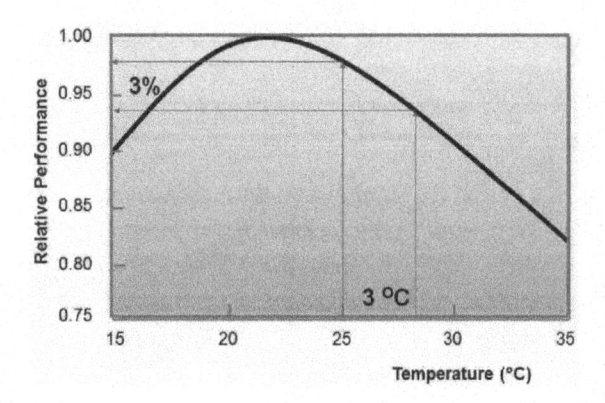

Figure 2.1. Relative performance of workers is depending e.g. the air temperature of the space. [66]

As an example, a reduction of indoor air temperatures by 1 °C can roughly increase the performance of office work by 1% (valid in temperatures above 22 oC) (Fig.2.1.). Doubling the outdoor air supply rate can reduce sick leave prevalence by 10 %, and increase office work by 1.5%. [70] With proper commissioning we can ensure that these benefits are realized.

In India, the ambient air quality does not meet ambient quality standards most of the time. To ensure good indoor air quality and to avoid the infiltration of polluted ambient air, it is highly necessary to have a tight building envelope and keep the entire building positively pressurized as compared to the ambient air. This can be guaranteed by focusing on the building's enclosure and balancing of ventilation system during commissioning. The high ultra-fine particulate levels in the ambient air are one of the major challenges, as they are 5-15 times higher in India [24] than WHO recommended levels [74]. As the ultra-fine particulates are carcinogenic for humans [37], a high particulate level leads to increased long-term mortality risk.

A part of the commissioning process is to ensure that maintenance personnel are trained to understand and operate all the technical systems in the building; their ability to handle user complaints becomes better and therefore user satisfaction improves.

2.2. Increased value for owners and investors with minor additional cost

Owners and investors benefit financially from well-commissioned, sustainable buildings. Performance of HVAC and other mechanical systems impacts indoor environment quality (IEQ), energy consumption and maintenance requirements. Proper commissioning of HVAC-systems affects both Relative performance of workers is the performance of the building (e.g. energy use and indoor air temperature) and the life cycle costs.

Both, building performance and LCC, have an impact on the building's value (Fig.2.2.). For owners, improvements can result in increased property value such as:

- Reduced life cycle costs;

- Extended building and equipment life span;

- Longer tenant occupancy and lease renewals;

- Reduced churn costs;

- Reduced insurance costs;

- Reduced liability risks;

- Lower interest rate of loan;

- Brand value.

In developed countries, the market value of a building and its environmental performance are directly correlated. As awareness about environmental issues grows worldwide and international and national authorities strengthen related regulations, the gap between the value of a well performing building as compared to a conventional buildings is only set to increase. (Fig.2.3.)

Lower energy and operations costs and higher rent income increase the net operating income of a building. A higher operating income discounted as a cash flow increases the market value of a building.

In owner-occupied buildings, the benefit of better productivity of workers benefits the owner directly, but in a speculative office market (where developers begin construction without locking down the leases) the higher rent is an instrument to transfer some of the user benefits to the owner who then has the ability to invest for better indoor environment. (Fig. 2.4.)

According to latest research, the impact of sustainability on a property's initial cost is typically

less than 3 per cent [30]. In many cases, more efficient performance can be achieved with no additional cost. Comprehensive commissioning is one of the 'additional' costs. According to the Department of Energy (DOE) in the United States of America (US), typical commissioning cost of a building is between 0.5-1.5 per cent of construction costs. [47]

A case study of 409 buildings in California 2009 [44] showed that the average commissioning cost of a new construction was 1.16 USD/ft2 and for existing buildings, 0.30 USD/ft2. Average energy saving was 13-15%. Cost saving was in average 0.18 USD/ft2, year in new construction and 0.29 USD/ft2 year in existing buildings.

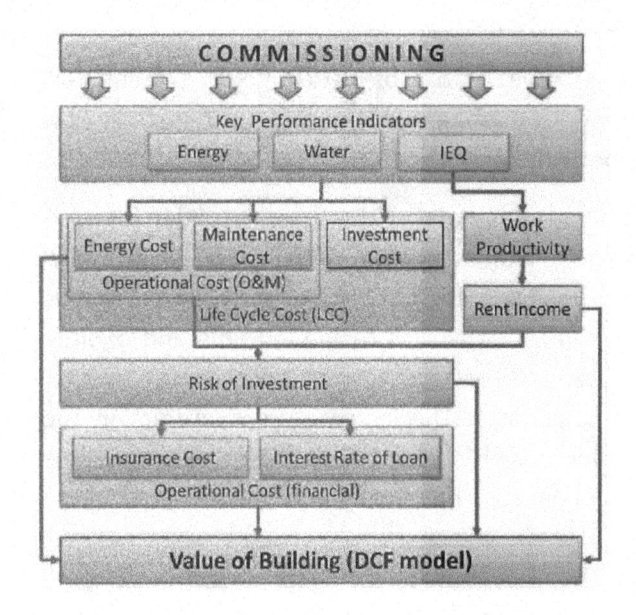

Figure 2.3. Building performance is strongly linked to a proper commissioning. Good performance indicators ensure low building's operational costs and high rent income. Furthermore, they influence the risk of investment that is also linked to an insurance cost and an interest rate of a loan. All these parameters together specify the current market value of a building.

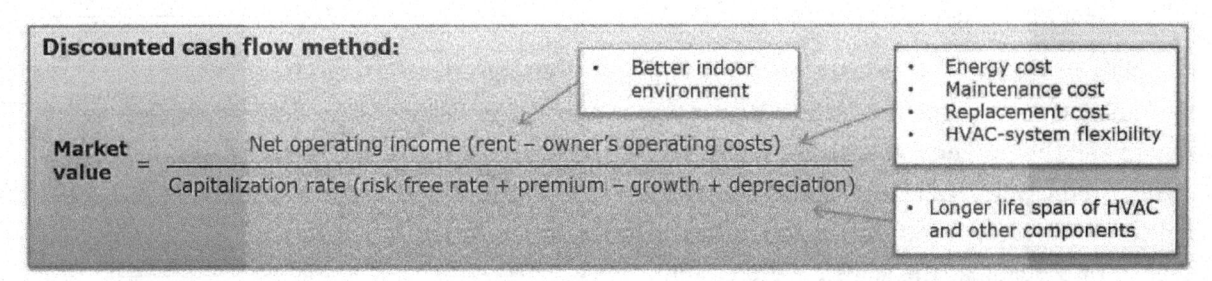

Figure 2.2. Building's market value that is calculated using discounted cash flow method is dependent on the performance of the HVAC system. [68]

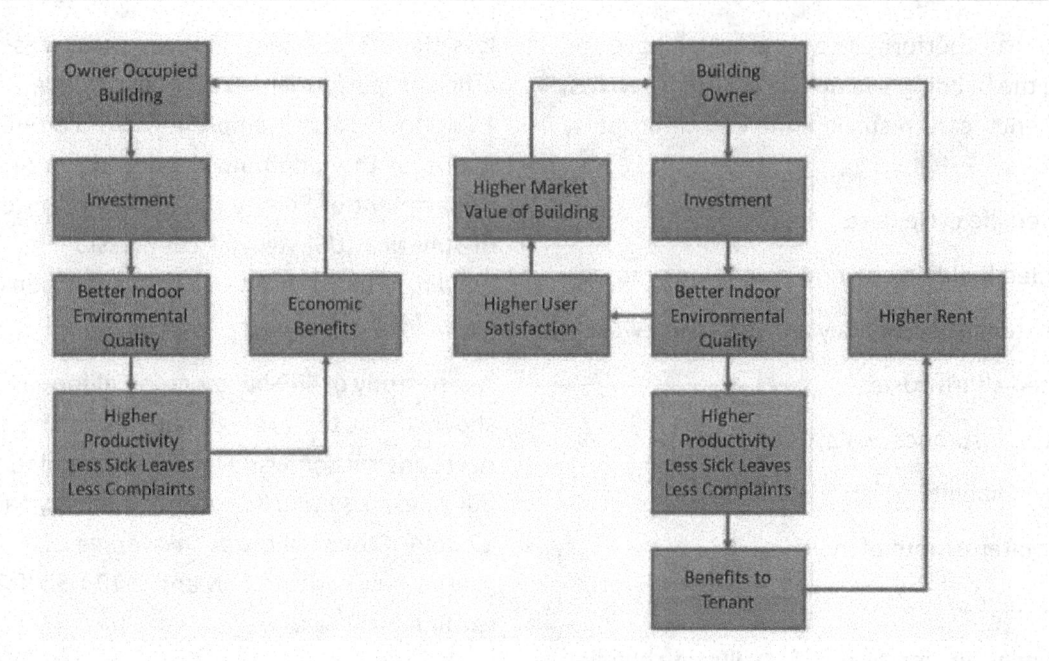

Figure 2.4. Economic benefit of better indoor environment and productive workers can be transferred to the owner by paying a higher rent. [70]

This results into a payback time of 1 month - 5 years in existing buildings and 1-10 years in new constructions.

Complete and useful O&M documentation is a basis of good maintenance. Well-trained O&M personnel can maintain the building properly, do preventive maintenance and all repairs on-time and operate systems in a way that comfort is achieved and energy is not wasted. This increases user satisfaction and reduces both the annual operation cost and long-term refurbishment debt.

2.3. Developer can sell a well performing building faster with higher profit

A developer's main focus is to develop a building that fulfils the targets set for performance, cost and time as-well-as to be able to sell the property fast with a premium. Focus on commissioning ensures that the developer can get what has been ordered.

As the performance has an impact on the value of the building (see chapter 2.2), the developer is likely to sell the property at a higher price (Tab. 2.1). IGBC Green, GRIHA, LEED, WELL or other green building certificates also contribute to the value and sales price of a building. The sales price premium, according the various studies in LEED certified buildings in the US, is 11-25 per cent. [67]

Table 2.1. Summary of green office value studies in the United States of America [71].

	Rental Premium	Sales Price Premium	Vacancy Rate Premium
Furest & McAllister (2011)	Energy Star 4%	Energy Star 26%	Energy Star 1.3%
	LEED 5%	LEED 25%	LEED – no premium
Eichhohz et. al. (AER)	Energy Star 3.3%	Energy Star 19%	"Effective rent": 7% premium overall
	LEED: 5.2%	LEED: 11%	15%
Eichholtz et. al. (RICS)	Energy Star: 2.1%	Energy star: 13%	"Effective rent": 6-7% Premium overall
	LEED 5.8%	LEED 11%	NA
Pivo & Fischer	2.70%	8.5%	NA
Wiley et. al. (2010)	Energy Star: 7-9%	NA	Energy Star: 10-11%
	LEED: 15-17%	LEED: 16-18%	NA
Miller et. al. (2008)	9%	None	2-4%

Proper commissioning also means better documentation during the entire construction process. This supports the technical due diligence during the sales process. There is less hassle during the construction project and unnecessary delays during start-up and handover can be avoided. This means shorter building handing-over period with a lowering of additional costs during construction and easier time management. Lower cost and on-time construction ensure s a higher profit for the developer.

2.4. A satisfied customer brings more business to the entire design and installation team

The Commissioning process supports the design and installation teams to focus on the total building performance, from the target setting to verification of performance at the end of the construction process. Verified performance helps them to demonstrate that the targets set in the Owner's Project Requirements and Basis of Design are met.

Good commissioning process minimizes the need and effect of changes both in the design and construction phase. The project will be on-time and start-up process is better managed due to proper planning. The building handing-over period will inevitably be shorter and with less unplanned post-occupancy visits to the site.

As the project meets the Owner's Project Requirements, the end client is happier and likely to give future design work to the same design team. Well-executed projects enhance the company brand to bring more demanding work in the future. Value-added quality construction is key to proper lifetime performance.

2.5. Products can deliver all benefits promised by manufacturer

Today advanced technologies to improve the indoor environment quality and energy efficiency of a building are available. But these benefits can only be achieved if the commissioning process is well managed and new technology is fully understood and implemented by installation and O&M personnel. Therefore, proper commissioning planning and cooperation of the entire design and installation team in vital for manufacturers to deliver results and satisfy the customer. Better planning also reduces change of orders and additional claims, creates fewer project delays, leading to less post-occupancy corrective work for manufacturer.

2.6. Proper system performance reduces energy use and environmental load.

The main environmental load of commercial buildings is the carbon emission related to the lifetime energy use. Efficient building performance can be met only if systems are performing in optimum operation range and delivering the desired indoor environment conditions to the occupants. Incorrect set values in terms of indoor environment conditions, wrong operating time or systems and oversized equipment result in inefficient operation and unnecessary high energy use.

Energy Performance Index (EPI) range for various commercial building types in India: 'Public sector office buildings' have EPI of about 100 kWh/m², year and 'one-shift commercial office buildings' have been benchmarked at 150 kWh/m², year. By integrating energy conservation mentioned in the Energy Conservation Building Code (ECBC), it is expected that the overall energy requirement for commercial buildings can be reduced by 30-40%. The ECBC provides both mandatory and prescriptive requirements for five building components: Building Envelope, Heating, Ventilation and Air Conditioning (HVAC), Service, Water, Heating and Pumping, Lighting and Electrical Power. [21]

All these elements are in focus during the commissioning process. But initial commissioning cannot guarantee good building performance alone and therefore continuous commissioning over a building's life cycle is vital. The energy performance of a building needs to be one of the key performance indicators of the building and should be reviewed annually. Regular inspection of the building and its mechanical systems are also required.

Business Case of Investments in Improved Building Operation in India

An office building's construction costs are typically 3,500 INR/ft$_2$ (540 €/m^2) in India. Typical HVAC cost is 250 INR/ft^2 (38 €/m^2), but in this case the developer invests 50% more i.e. 375 INR/ft^2 (57 €/m^2). The HVAC designer will get an additional fee of 3 INR/ft^2 (0.5 €/m^2) to ensure high-quality design. The developer has also decided to employ experienced commissioning provider and invest 5 INR/ft^2 (0.8 €/m^2) towards the initial third-party commissioning instead of the typical fee of 2 INR/ft^2 (0.3 €/m^2). Therefore, the additional investment will be 131 INR/ft^2 (20 €/m^2) i.e. 3.7% increase in construction cost.

Annual energy cost in a typical design case would be 120 INR/ft^2 (18 €/m^2). Due to additional investment on good design and efficient system operation, the energy use will be reduced by 20% i.e. 24 INR/ft^2 (3.7 €/m^2). The client will also put more effort on system maintenance, and therefore maintenance cost will increase from 80 INR/ft^2 (12 €/m^2) to 85 INR/ft^2 (13 €/m^2).

As the indoor environmental quality will be better, the productivity of workers will increase by 2% (10 min/day). Total annual salary cost (CTC) of each person working in the building is in average 5,00,000 INR (7,100 €) and each person occupies 70 ft^2 (6.5 m^2) of floor space. Annual salary cost saving will be 140 INR/ft^2 (13 €/m^2).

Typical office rent in Delhi is 100 INR/ft^2 (15.4 €/m^2) per month. As the building now provides better indoor environmental conditions for users, tenants are paying 2% higher rent (2 INR/ft^2,month (0.3 €/m^2,month)) i.e. 24 INR/ft^2 (3.6 €/m^2) annually.

The value of the office building is calculated using the discounted cash flow (DCF) method. Due to the higher rent and lower operating costs, the value of the building will increase from 14,900 INR/ft^2 (2,300 €/m^2) to 15,600 INR/ft^2 (2,400 €/m^2) i.e. +4.6%. The higher value will be shared between the developer and the owner. Therefore, building will be sold at 15,400 INR/ft^2 (2,350 €/m^2).

As the investments during the construction phase are higher than in a typical project, they are often cut out in a 'cost driven' project. However, in a 'quality driven' project, each stakeholder understands the holistic benefits. In this case, each stakeholder benefits as follows:

User:

- *increased productivity - higher rent* .... **+116 INR/ft^2,year (9.4 €/m^2,year)**

Owner:

- *higher rent income + lower energy cost - higher maintenance cost* **+43 INR/ft2,year (6.3 €/m2,year)**
- *higher value of building* .... **+700 INR/ft2 (100 €/m2)**

Developer:

- *higher sales price - higher investment* .... **+219 INR/ft^2 (30 €/m^2)**

Design team:

- *increased design fee* .... **+3 INR/ft2 (0.5 €/m2)**

HVAC-contractor and manufacturer:

- *higher equipment value* .... **+125 INR/ft2 (19 €/m2)**

Commissioning team:

- *Increased fee* .... **+3 INR/ft2 (0.5 €/m2)**

3. INITIAL COMMISSIONING

3.1. Objectives and goals of commissioning

Commissioning is a 'support process' to assist design, installation, start-up and maintenance processes to arrive, from the owner's project requirements, to the desired results i.e. a well-performing, sustainable and high value building (Fig. 3.1.). Any commissioning process translates the Owner's Project Requirements (OPR) to the specifications, which can be understood and realized by the designers and contractors.

Commissioning (Cx) will start with the development of the OPR and setting the critical Key Performance Indicators (KPIs) that are diligently monitored during commissioning operations. Its planning is important in order to communicate what to do, when and who is responsible for each activity during commissioning. Functional performance testing procedures enable verification of system operation as per the OPR.

Top-down approach consists of 'going down' from the 'system' level to the component 'level',

passing through some sub-systems. The entire system's functional performances are first verified and the inquiry is extended to lower levels, through analysis and synthesis of data on system parameters. The final goal is not to verify whether a component is 'good' or 'bad' in itself, but to check if it is correctly integrated in the system being considered.

The 'bottom-up' approach consists of starting from an installation of basic component and going up progressively to cover the entire system operation. This approach is appropriate for initial commissioning, in order to follow the construction (not to arrive too late) and allow a safer identification of local faults.

System O&M and user manuals enable the operator comparing the prevailing operating parameters and the parameters recorded during the commissioning. Regular interval reporting structure and content specified during the initial commissioning, enable the operator and the building owner to check its operation continuously to fulfil the set requirements.

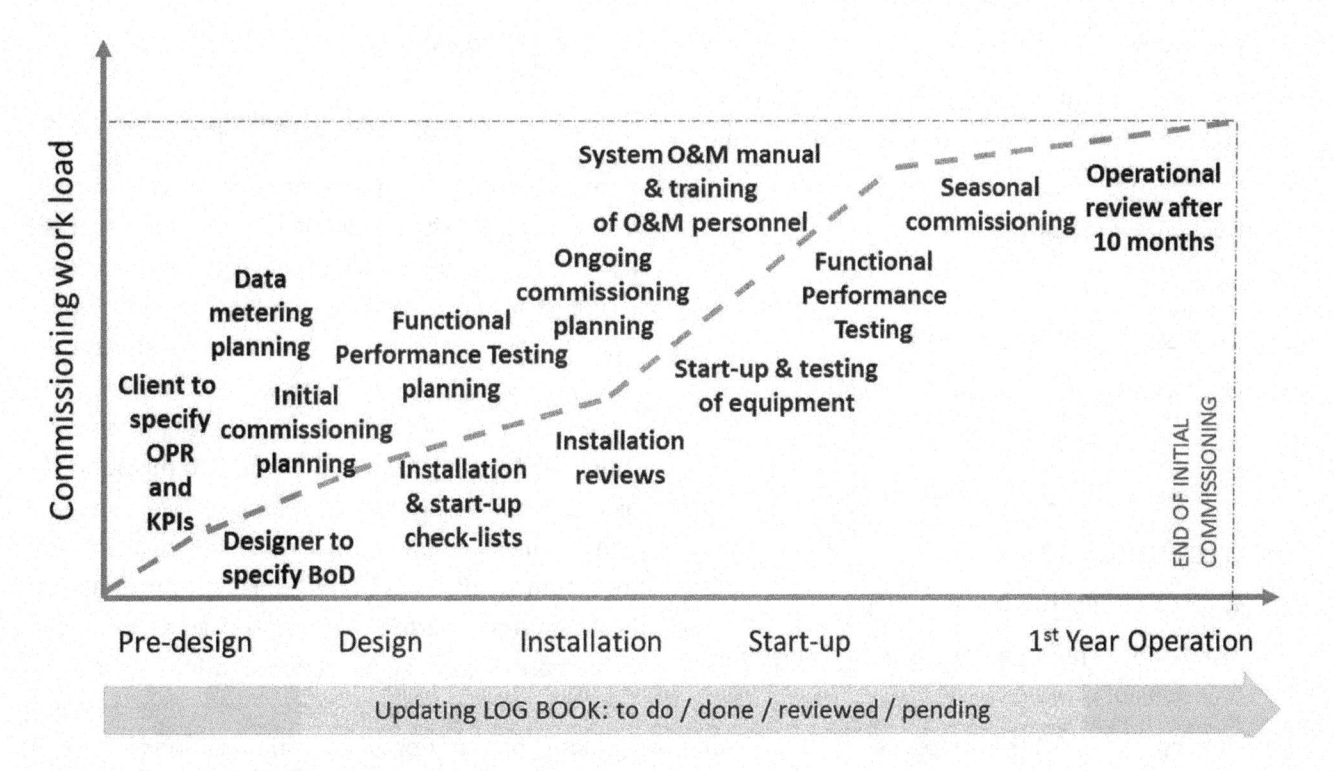

Figure 3.1. Activities during the initial and seasonal commissioning.

Commissioning Provider together with the contractors executes the functional performance tests (FPT) for a system operation. An active test will assess performance by analyzing data obtained from systems that are subjected to changes in operational conditions.

Those systems that are dependent on external weather or internal occupancy pattern need to be commissioned in each occupancy load and season (summer, monsoon, winter). Initial and seasonal commissioning should lead to the continuous commissioning over the building's life cycle to resolve operating problems, improve comfort, and optimize energy use.

3.2. Systems to be commissioned

Commissioning is required for all operational systems that have an impact on indoor environment conditions, user safety, energy use of building or water consumption. Therefore, the following services need to be addressed when planning and implementing commissioning activities in the building:

Mechanical ventilation system:

- Ductwork cleanliness, tightness and insulation;
- Dedicated outdoor air units (DOAS), treated fresh air units (TFA) and air handling units (AHU);
- Exhaust and supply air fans;
- Volume control dampers (VCD), variable air volume (VAV) dampers and constant air volume dampers (CAV);
- Ductwork balancing;
- Room air diffusion.

Natural ventilation system:

- Operational windows and their controls;
- Façade louvers to control natural ventilation;
- Exhaust and supply fans.

High and low side cooling:

- Chillers, cooling towers and pumps;
- Installation, flushing, pressure testing and balancing of pipework;
- Unitary products like window AC, split AC, and variable refrigerant flow (VRF) system;

- Room cooling units like floor cooling, radiant cooling, chilled ceilings and chilled beams;
- Other room air condition units like precision air conditioning units (PAC), in-row coolers (IRC) and door coolers in data centres;
- Shut-off, balancing and control valves.

Heating systems:

- Boilers and heat exchangers;
- Installation, flushing, pressure testing and balancing of pipework;
- Room heating units like radiators and floor heating;
- Shut-off, balancing and control valves.

Building management system (BMS):

- Measurement sensors;
- Energy (electricity) & BTU (thermal energy) meters;
- Operation principles and operation parameters of all systems through BMS system;
- Room conditions like air temperature, RH, CO_2 and $PM_{2.5}$.

Ventilation fire safety systems:

- Fire and smoke dampers;
- Smoke exhaust fans;
- Staircase pressurization.

3.3. Initial commissioning activities

Initial commissioning covers the basic commissioning required to satisfy a building's specifications. It is a proof of the building's capability to deliver the required performance in terms of system operation, indoor environmental quality, energy efficiency, safety and sustainability. The National Conference on Building Commissioning in the US has established an official definition of 'Total Building Commissioning' as a 'systematic process of assuring by verification and documentation, from the design phase to a minimum of one year after construction, that all facility systems perform interactively in accordance with the design documentation and intent, and in accordance with the owner's operational needs, including preparation of operation personnel.' [72]

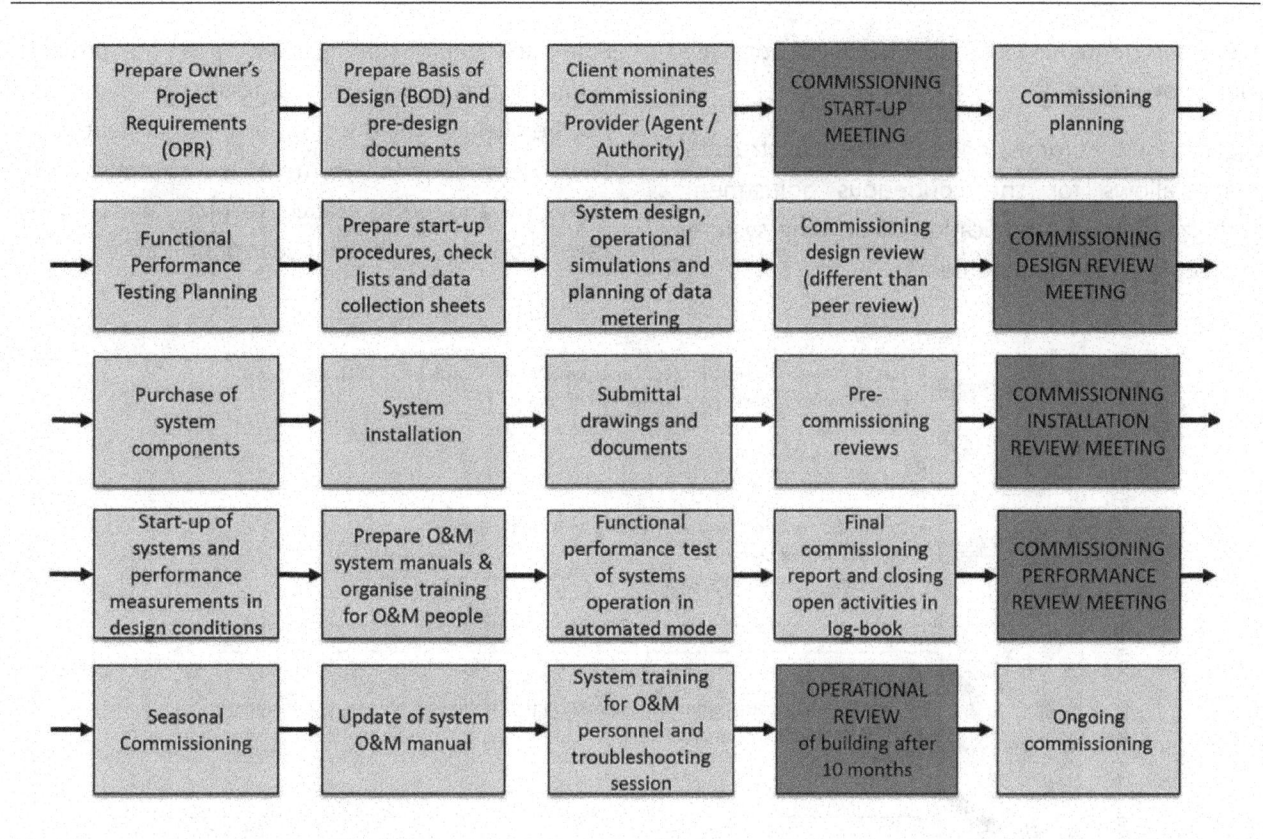

Figure 3.2. Initial commissioning activities.

Initial commissioning (Fig. 3.2. and Fig. 3.3.) starts in the planning phase. Consideration for commissioning is critical at this phase of a project as commissioning activities need to be incorporated within the project delivery process and commissioning ought to be part of project scope, budget and time schedule.

The commissioning scope specifies the systems to be commissioned and outlines the activities. For each project, the commissioning purpose and scope should be clearly defined in the various contracts:

- Project management;
- Designers (architectural, MEP);
- Commissioning provider (authority/agent);
- Contractors (MEP).

The project budget needs to be sufficient to support the commissioning activities and the construction schedule should provide enough time to accomplish commissioning activities. Budget is impacted by the scope of work. Therefore, it is recommended to use professional help to determine how many of each of the systems are evaluated and verified.

Commissioning is a quality process that includes adequate target setting, design and installation reviews, system start-up, functional performance testing and training to ensure the desired system performance. The owner shall develop the owner's project requirements (OPR) that will establish the benchmarks for performance later in the commissioning process.

Design phase commissioning activities serve to ensure that the OPR is sufficiently defined and adequately addressed in the contract documents. Commissioning plans as well as start-up tests and procedures are developed to verify the performance of systems and assemblies and are incorporated into the contract documents.

The commissioning team collaborates during the construction phase to verify that systems are installed correctly and to validate that the Owner's Project Requirements (OPR) are achieved. The commissioning team needs to declare that technical systems will function according to user expectations. Testing and documentation will also provide important benchmarks and baseline data for future on-going commissioning of the building.

Systems and equipment will tend to drift from their as-installed conditions over time. In addition, the needs and demands of building occupants typically change over time.

The seasonal commissioning in post-construction phase allows for the continuous adjustment, optimization and modification of building systems to meet specified requirements.

Full-load commissioning may not be possible prior to occupancy because external loads may not be available at the time or internal heat gains are absent. Therefore some of the commissioning activities and performance testing needs to be carried out during the first year of operation.

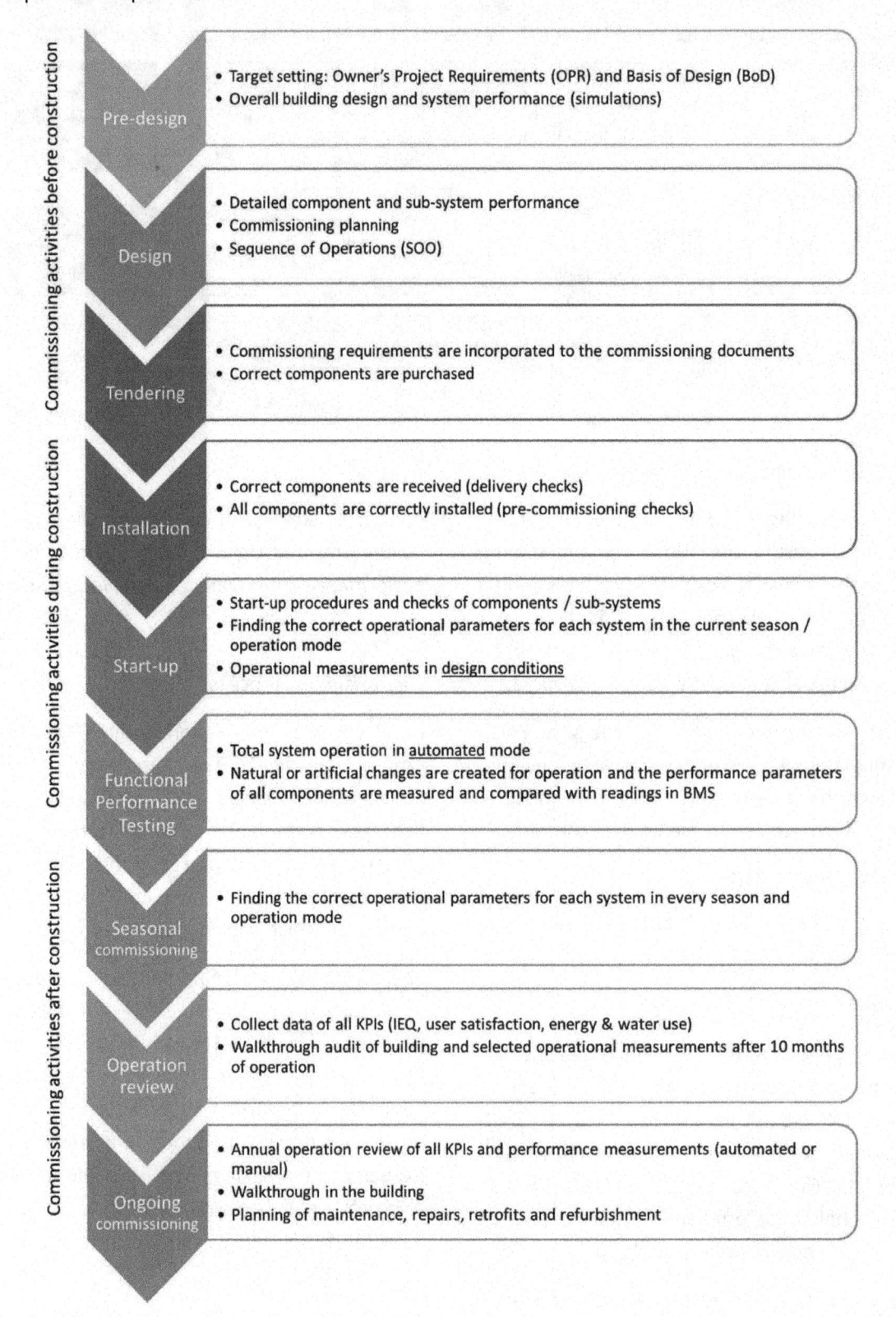

Figure 3.3. There are several key commissioning and building performance related activities in each phase of construction and also after the building is occupied.

3.3.1. Owner's project requirements (OPR) and key performance indicators (KPI)

The owner shall develop and distribute the Owner's Project Requirements (OPR) including the Key Performance Indicators (KPI) to measure building performance, economy and user satisfaction in the planning phase.

OPR defines the functional requirements of a project and the expectations of the building's use and operation. The OPR is developed with significant owner input and approval, but typically Commissioning Provider, if already nominated, assists the owner in identifying the facility's requirements.

An effective OPR is a living document that incorporates inputs from the owner, design team, operation and maintenance staff and end users of the building early in the project and is updated throughout the project. Targets need to be concrete, verifiable and/or measurable in order to use them as a benchmark of operation later in the commissioning process. The owner is responsible for updating the document later in case there are any changes.

A good OPR includes the following information:

- Performance targets of the building:
- Primary purpose of the building;
- Space program, number of occupants and typical usage profiles;
- Flexibility to change floor layout and the use of spaces and/or building;
- Training required for operating and maintaining building.

- Targets for indoor environmental quality (e.g. based on the ISHRAE IEQ standard 10001) [40]:
- Thermal comfort;
- Indoor air quality (IAQ);
- Lighting;
- Acoustics;
- User satisfaction.
- Health and safety of occupants:
- Security (door access, exits etc.);
- Water safety;
- Emergency lighting;
- Fire and smoke safety strategy.
- Data management:
- Phone connectivity;
- Wireless business data network;
- Wireless IoT data network.
- Key environmental performance targets:
- Energy Performance Index (EPI);
- Primary energy consumption (takes into account the losses in energy production and supply);
- Existing and possible future means of energy supply;
- Specific water-use of building;
- Water use reduction potential by tenants.
- Lifetime greenhouse gas emissions from cradle to grave including embodied energy of construction materials, transportation of materials, construction work, lifetime energy use, maintenance, retrofits, repairs, recycling of materials in the end of lifecycle and demolishing of building;
- Operational greenhouse gas emissions including, e.g. annual energy use and commuting of occupants.
- Waste (hazardous/non-hazardous).
- Targets for on-going commissioning & continuous monitoring.

Key Performance indicators should measure the sustainability through the 'Triple Bottom Line' approach (Fig. 3.4.), i.e. economic, social and environmental targets to ensure the wellbeing of people occupying the building, making sure that

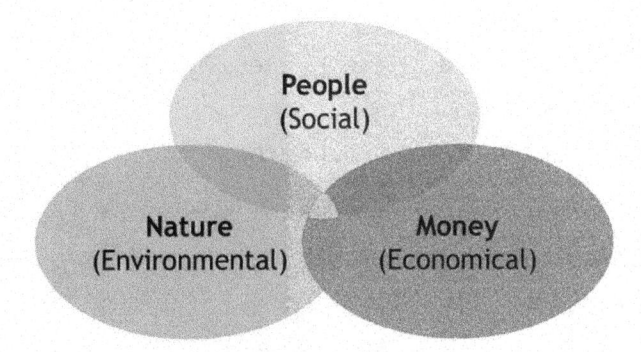

Figure 3.4. Key performance indicators should include measures from all three categories, people – nature – money, to ensure the sustainable building operation.

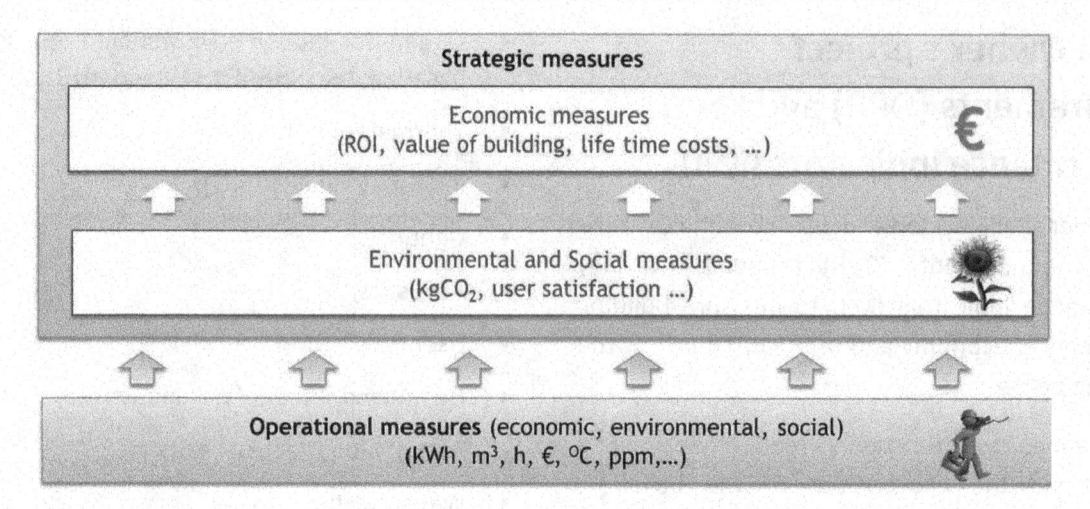

Figure 3.5. Operational measures are required to specify the more important strategic measures. OPR should set targets for both groups. [69]

focus is not only on initial cost but also on Life Cycle Cost (LCC) and ensuring that the building does not unnecessarily increase the environmental impact.

It is important to identify strategic measures without falling back on things that are easy to measure. (Fig. 3.5.) Strategic measures of building operation can be as an example:

- Value of building;

- Life cycle cost (LCC);

- Return on Investment (ROI);

- Amount of green lease and service contracts;

- User satisfaction index;

- Energy Performance Index EPI (e.g. according Energy Conservation Building Code)

- Carbon footprints (lifetime/energy use/commuting);

- Water footprint;

- Environmental/green building certification, e.g. IGBC Green, GRIHA, LEED or WELL.

Most of the strategic measures require operational data to specify and calculate them. Typical operational measures of a building are:

- Indoor Environmental Quality Class (according ISHRAE standard) or IEQ parameters separately:

  - Room air temperature (dry-bulb/operative) and deviation from target value;

  - Respirable Suspended Particulate Matter level (RSPM);

- Total Volatile Organic Compounds (TVOC) and Formaldehyde (CH2O) level;

- Carbon dioxide (CO_2) and Carbon monoxide (CO) levels;

- Total Microbial Count;

- Outdoor air gases: Sulphur dioxide (SO_2), Nitrogen Dioxide (NO_2), and Ozone (O_3);

- Sound levels in A-weighted Sound Pressure Level (SPLA) or based on NR- or NC-criteria.

- Advanced sound criteria like Speech Transmission Index (STI), Spatial Decay Rate, Distraction Distance, Privacy Distance and Reverberation Time;

- Illumination levels when artificial and day-lighting are used;

- Day-lighting factor;

- Uniformity of illumination and United Glare Rating (UGR);

- Requirements for circadian lighting, e.g. specifying Equivalent Melanopic Lux (EML) levels;

- Use of rare or energy intensive construction materials;

- Soil transportation during construction;

- Amount of landfill waste during construction;

- Waste management targets during operation;

- Water and energy use and cost;

- Energy use of an empty building;

- Renewable energy use;

- Operating cost.

3.3.2. Basis of Design and Pre-design

The design team shall develop and update the Basis of Design (BoD) document for the systems to be commissioned prior to approval of contractor submittals of any commissioned equipment or systems.

The following items should be considered in Basis of Design documents:

- Primary design assumptions:
  - Space use;
  - Redundancy;
  - Diversity;
  - Climatic design conditions;
  - Space zoning;
  - Occupancy;
  - Operations;
  - Space environmental requirements;
  - Standards and codes.
- Technical performance criteria for:
  - HVAC&R-systems;
  - Lighting;
  - Water management including e.g. cooling tower replacement, domestic hot water (DHW) and irrigation;
  - Fire and smoke safety;
  - On-site energy production;
  - Onsite waste and sewage management.
- Equipment and system performance targets:
  - Adaptability of mechanical systems;
  - Maintainability;
  - Inspection access;
  - Reliability;
  - Required total service life of components;
  - Schedule of replacements;
  - Specific efficiency targets for key components and systems.

The Commissioning team reviews the BoD and other pre-design documents and determines whether they are consistent with and support the Owner's Project Requirements.

In terms of user comfort and energy efficiency, it is also important to simulate building operation and typical spaces in pre-design phase. This way the entire design team can be confident that selected technical solutions can fulfil the performance targets of the project (set in OPR and BoD). Typical simulations in pre-design phase are daylight and lighting simulations, annual energy use and room air temperature simulations as-well-as use of Computational Fluid Dynamics (CFD) tool to simulate comfort conditions and air movements in the building.

3.3.3. Nomination of commissioning provider

The owner shall designate a Commissioning Provider (or Authority or Agent) for the project in pre-design phase or early design phase. The Cx provider reports directly to the owner and is in charge of the commissioning process.

Cx provider makes the final recommendations to the owner regarding functional performance of the commissioned building systems. It is recommended, especially in the larger projects, that the Cx provider is an independent, third party company having no other role in the project. This is to ensure that the Cx provider can be an objective and independent advocate of the owner.

Commissioning provider should have knowledge of:

- Building systems;
- The physical principles of building systems performance;
- Building systems start-up, balancing, functional testing and troubleshooting;
- Operation and maintenance procedures;
- Testing and measurements;
- The building design and construction process;
- Commissioning process.

More about commissioning provider's role and responsibilities would be covered in chapter 6. Commissioning Management.

3.3.4. Commissioning planning

Commissioning planning is one of the key activities during the commissioning process. It consists of the following plans, checklists and procedures (Fig.3.6.):

- Initial commissioning plan;

- Product and sub-system related procedures and checklists (delivery, pre-commissioning, start-up and operation) for each operational component;

- Functional performance testing plan (entire system operation);

- Seasonal commissioning plan;

- On-going commissioning plan.

Each project needs to have a written Commissioning Plan that is updated as the project advances from conception to completion. The Commissioning Plan establishes the framework for how commissioning will be managed on a given project. All information in the Commissioning Plan must be project specific and having the following content:

- Project definition and functional description of the building;

- Project site address;

- Project team member directory;

- Communication structure/meetings;

- Project schedule;

- Objective and goals of commissioning;

- Systems to be commissioned;

- Commissioning provider's role;

- Commissioning process activities;

- Expected work and deliverables by each party.

The commissioning plan shall be provided to all designers and contractors as a supplement to the construction documents.

Product and sub-system related procedures and checklists need to be done for each operational component and they are provided either by the client as standard in-house documents, or developed by the commissioning team in each project specifically. (Fig. 3.7.)

Delivery checklists are used to check all major operational equipment once they arrive at the building site.

The following equipment needs to be checked:

- Dedicated outdoor air units, treated fresh air units and air handling units;

- Exhaust and supply fans;

- Variable Air Volume (VAV) dampers;

- Fire and smoke dampers;

- Chillers;

- Cooling towers;

- Pumps;

- Heating boilers;

- Heat exchangers

- Unitary products like split AC, cassette units, heat pumps and variable refrigerant flow (VRF) system (indoor and outdoor units);

- Precision air condition units;

- In-row coolers and door coolers;

- Fan coil units;

Commissioning Plan	Procedure descriptions	Contractor check lists for each operational component	Operation data collection sheets for each operational component	Functional Performance Testing Plan	Seasonal Commissioning Plan	On-going Commissioning Plan
• Project team; • Communication structure; • Project schedule; • Systems to be commissioned; • Commissioning activities; • Expected work and deliverables. of each party.	• Describes how different start-up activities or measurements shall be carried out (15-20 procedure descriptions required depending on the project)	• Delivery • Pre-commissioning • Start-up. (30-50 check lists required depending on the project)	• To record operation data in design conditions (15-30 data collection sheets required depending on the project)	• Monitoring KPIs • Data collection sheets to record operational data in automated mode when system operation mode is changed (3-5 systems)	• Tests and data collection required during the first year of operation in different seasons	• Annual actions to evaluate system performance , set targets for operation and process to plan repairs, retrofits and refurbishment.

Figure 3.6. Several plans, procedure descriptions, check list and data collection templates need to be prepared during the commissioning planning.

- Radiant cooling panels and manifolds;
- Chilled beams;
- Balancing and control valves;
- Measurement sensors;
- Energy and BTU meters;
- Weather station.

Typical items to focus on are product specification and nameplate as-well-as completeness of components and accessories as per specification. It is also important to visually check that there are no transportation damages. In case the component is not correct or there are transportation damages, the contractor shall inform the supplier immediately to get the replacement delivered to the site as soon as possible.

Once the contractor signs the delivery checklist, he approves that the delivery is complete. A copy of the Delivery Checklist shall be given to the supplier.

Pre-commissioning audit is required to ensure that the system is ready for start-up. Checklists are required for all operational components (mentioned above) as-well-as for ductwork, pipework and BMS as a whole.

Typical items to be checked are for example:

- Cleanliness of all components, ducts and pipes as well as AHU rooms;
- Installation of all components (completeness, quality, etc.);
- Pressure testing of pipework;
- Leakage testing of ductwork;
- Water quality of pipe works;
- Connections (air, water and electrical).

Detailed pre-commissioning checks of each component are presented later in this book. Also if required, ensure that factory-made performance testing results are available during the start-up process. After pre-commissioning checks, systems are ready to be started.

Start-up procedures and checklists describe the method how each start-up activity is carried out and what kind of data needs to be verified. Start-up tests validate that the component or sub-system is ready for automatic operation in accordance with the manufacturer's requirements.

The following components and systems need to go through the start-up process:

- Dedicated outdoor air units (DOAS), treated fresh air units (TFA) and air handling units (AHU) with Variable Frequency Drivers (VFD);

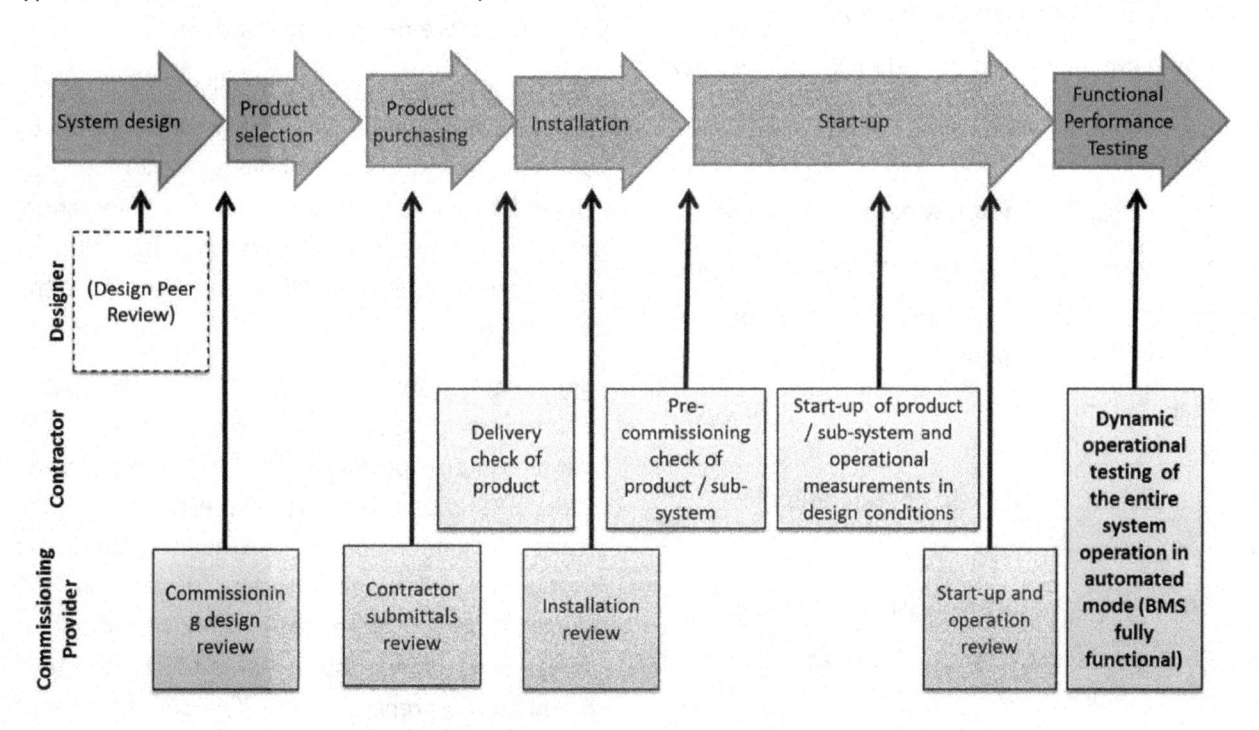

Figure 3.7. The Contractor needs four different checklists to commission each component and in addition the commissioning provider needs another set of checklists for review.

- Exhaust and supply fans;
- Ductwork (balancing);
- Chillers;
- Cooling towers;
- Pumps;
- Boilers and heat exchangers;
- Pipework (balancing);
- Variable refrigerant flow (VRF) system;
- Heat pump system;
- Precision air conditioning units (PAC);
- In-row and door coolers;
- Building management system (BMS);
- Fire and smoke management systems.

Operation checks and measurements focus on the operation of the equipment or system in design conditions. Following operation measurements as an example need to be carried out:

- Dedicated outdoor air units, treated fresh air units and air handling units (air flow rates, water flow rates, pressure differences across each component, air and water temperatures, RH);
- Variable Frequency Drivers (frequency with minimum and maximum air flow rate);
- VAV-dampers (minimum and maximum air flow rates and pressure loss);
- Exhaust fans (air flow rate, pressure);
- Room air diffusion (throw pattern, air volume);
- Fire dampers (open/closed, actuator operation);
- Smoke extract system (air flow rate of fan, damper and fan operation);
- Staircase pressurization (pressure difference, operation of fan);
- Operable windows (actuator operation);
- Chilled water quality;
- Chiller (temperatures, pressures, flow rates);
- Cooling tower (water quality, inlet and outlet temperature);
- Pump (head, flow rate);

- Boiler and heat exchanger (temperatures, water flow rate);
- Unitary products like split AC, cassette units, heat pumps and variable refrigerant flow (VRF) system (temperatures, pressure, vacuum, refrigerant filling);
- Chilled beam (chamber pressure/air flow rate, water flow rate, nozzle type);
- Radiant cooling system (surface temperature, water flow rate);
- Precision air conditioning system (AHU measurements and room air temperature during 24/48 h);
- In-row and door coolers (temperatures, flow rates);
- Energy and BTU meters (calibration);
- Building Management System (all measurement sensor readings).

The detailed measurement tables of the most common components are presented later in this book.

The contractor and/or commissioning provider shall also provide the list of measurement equipment that will be used during the commissioning and valid calibration certificates. After the operation tests, the contractor can be sure that all installed components are performing as designed.

Contractor submittal review, installation review and start-up & operational review checklists are used by commissioning provider to review the components selected by the contractor in terms of operation, installation as-well-as start-up and operation of equipment. Detailed checklists are presented in the appendix A.

Functional Performance Testing (FPT) (or integrated system testing (IST)) of systems focuses on the overall system performance in the automated mode. FPT is at the heart of the commissioning process. Commissioning provider plans the functional performance testing based on the owner's project requirements (OPR). At the end of the test Cx provider shall be able to sign the final commissioning report where it is confirmed that building and all its systems perform as per OPR and design documents.

During the start-up, each component has been tested separately. The functional performance test shall focus on the total system operation in automated mode. Therefore the FPT plan shall include the testing of all major systems in the building. After the functional performance test Cx provider confirms that key performance indicators are fulfilled in various operation conditions (occupancy, time of day, season and weather).

Typically the Functional Performance Testing plan focuses on the following areas:

- Key performance indicators:

- Thermal comfort like temperature, relative humidity, room air velocity;

- Indoor air quality like RSPM, CO2, HCHO, TVOC, CO, pressure difference between spaces and between indoor and outdoor air;

- Acoustic performance like background noise level (e.g. Equivalent Continuous Sound (Leq)) in the space;

- Lighting and daylight levels;

- Water use of building;

- Energy use of building.

- System operation (when changes in operation is created either naturally or artificially:

- Ventilation system;

- Cooling system;

- Heating system;

- Fire and smoke safety systems;

- Water and sewage system;

- Building management system.

Seasonal commissioning plan focuses on system operation under different load conditions. As different weather conditions cannot always be simulated during the initial system start-up, seasonal commissioning is needed during winter, summer and monsoon as well as under various occupancy conditions. This plan shall focus on the operation of:

- Ventilation system (treated fresh air units and dedicated outdoor air units);

- Cooling system (chillers and cooling towers);

- Heating system;

- Heat pump system.

On-going commissioning plan shall focus on follow-up of key performance indicators. Annual follow-up is needed in various systems like ventilation, cooling and building management system. It is also recommended to measure IAQ parameters annually if measurements are not part of continuous data metering. Building walk-through audit together with other data gives a basis for planning of annual repairs, retrofits and refurbishment. Also cleanliness and maintenance of systems need to be addressed. On-going commissioning is discussed more in detail in the chapter 5.

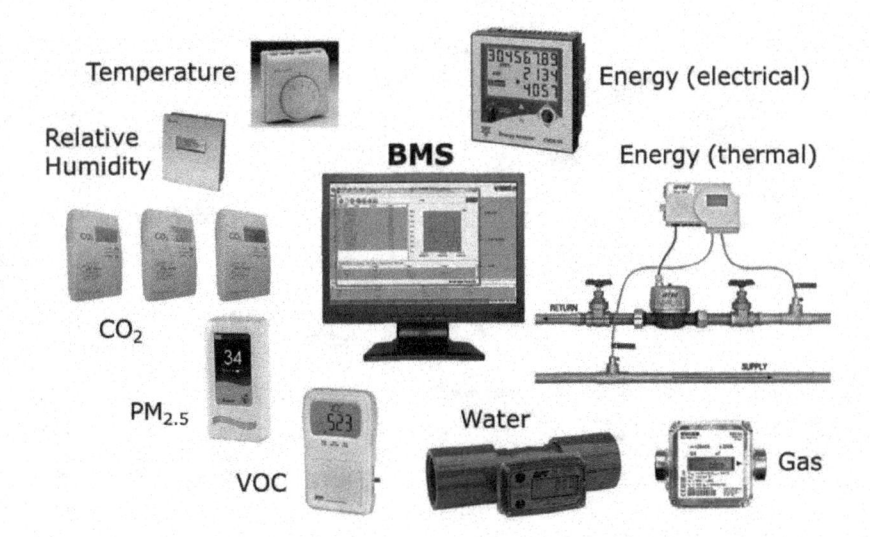

Figure 3.8. Data metering in a building should combine the indoor environmental quality, energy and water consumption as all these elements are important key performance indicators and used in the monitoring-based commissioning as well as in automated testing.

3.3.5. Planning of data metering

Intelligent buildings have changed traditional building automation by integrating Building Management Systems (BMS) with IT systems and introducing wider metering platforms to provide much more detailed picture of a building's operation. (Fig. 3.8.) The primary goal of BMS is to control the building operation & create alarms when required. However, we should expose information to a broader audience of stakeholders than BMS normally does. A data metering infrastructure combined with an Energy Management System (EMS) can provide opportunity for increased metering and monitoring accuracy, data logging, and enhanced circuit coverage. This is the fundamental foundation of an intelligent building and complements traditional BMS functionality.

Efficient data metering also enables 'monitoring-based initial and on-going commissioning' of a building. A key benefit of sub-metering is the ability to enhance the performance of buildings. The sub-metering data can be used as part of the fault detection and thereby identifying problems with installed equipment before occupants observe adverse effects in utility bills. Additionally, sub-metering data may identify nonstandard utility (e.g. energy and water) consumption as a result of system faults that may not be recognizable in the entire building's utility usage data. This detailed information enables more condition-based preventive maintenance in buildings, avoiding the higher costs typically incurred with deferred or unplanned maintenance.

Data metering helps to identify performance improvements and guides preventive maintenance. It creates trends of monthly and annual use of water and energy and thus, helps to identify the benefits of system upgrades. It also points out the systems (e.g. boilers, cooling towers) that may need repair or replacement or retrofit. Further it enables quick response to component failures, assuming the meters are linked to an EMS or a BMS. Data metering provides data to building occupants about the impacts of their behaviour on energy and water consumption. It allows lowering peak demand charges on electrical utility bills through virtual aggregation of different sub-meters.

It is important to consider how frequently (each second, minute, hour, day, week or month) each data measurement needs to be recorded or is one-time spot measurement good enough. This is very much dependent where and how data is used afterwards. Some of the measurement are continuous and are used to control the building operation, some are basis of payments to utility companies for building owners and tenants.

An important element over the life cycle of a building is to be able to follow-up the performance of the building and to measure the key performance indicators. Most of the occupied spaces already have

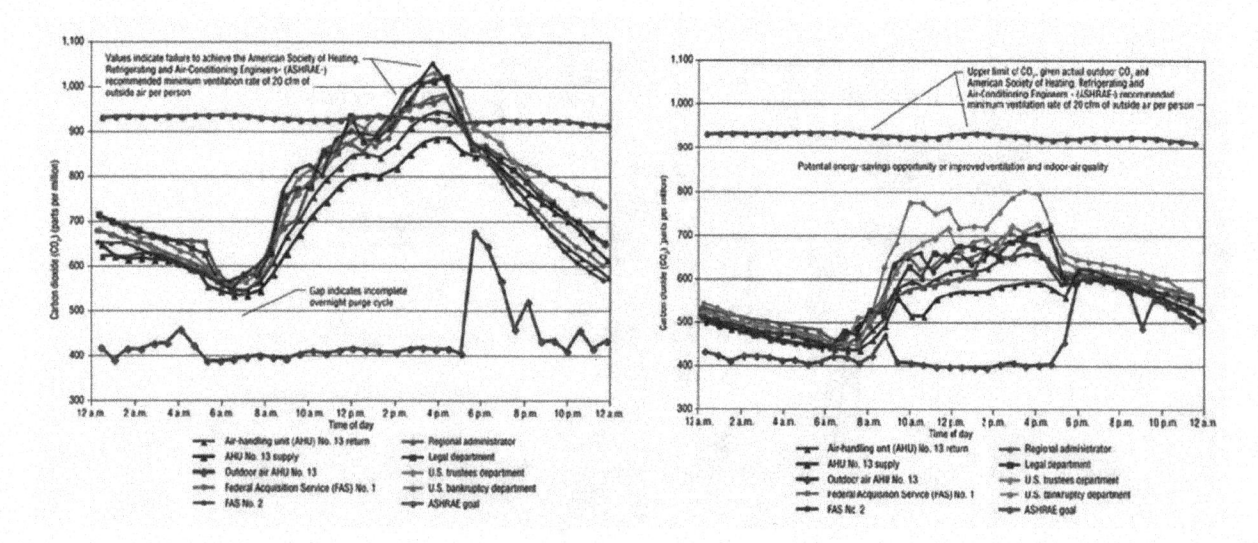

Figure 3.9. Examples of CO$_2$-measurement results and conclusions that can be made to control systems more efficiently in order to save energy and improve IAQ. [72]

the temperature, CO_2 and RH measurement. Measured results then need to be compared with the simulated ones. (Fig.3.11.) In addition to that, indoor air quality parameters like TVOC, CH_2O and RSPM should also be measured. However, these measurement sensors are still rather expensive to purchase and require continuous calibration. The pressure difference between the occupied spaces and ambient air is important to follow-up at least in each floor in order to keep the positive pressure in the entire building and thus avoid infiltration of polluted, hot and humid ambient air. Respectively, there are spaces and in general buildings in cold climate, where negative pressure needs to be maintained.

Several measurements are required in the air handling unit. It is important to have on-time data of air temperature and humidity after cooling coil to ensure that dehumidification process works as per design. Ambient air conditions, like temperature, humidity and CO_2 are important to know in order to decide the right operation parameters for cooling and energy recovery systems.

If return air is mixed with ambient air in the air handling unit, the CO_2-level of mixed air must be measured in order to ensure sufficient ambient air intake in all operation conditions. As the ambient air is often dirtier, warmer and more humid than return air from the building, it is more efficient to use as much return air as possible, but on the other hand we need to make sure that the CO2-level of supply air and in the spaces will not become too high in any operational conditions (Fig. 3.9.).

Energy metering is one of the key areas when designing a sustainable building. Although total building energy use is the key parameter to measure, sub-metering data is also required to better understand where the energy is used. Sub-metering can be either based on different areas of building (tenant or location based metering) or based on different subsystems or components (energy load specific metering).

Main utility metering typically comprises of electricity and gas metering at the point of entry of the supply to the building. These are used by the utility company for billing purposes and customers receive a monthly or quarterly bill that provides the cumulative energy consumption since the last bill. The meters provide limited information with which to manage the energy consumption and cost in a building.

Location based sub-metering is installed to separately measure the energy use of each floor or tenant. Such sub-metering helps the building owner to determine a tenant's energy use and to allocate their part of the energy bill. It is also important to measure the areas with highest electricity use, e.g. data centres in building.

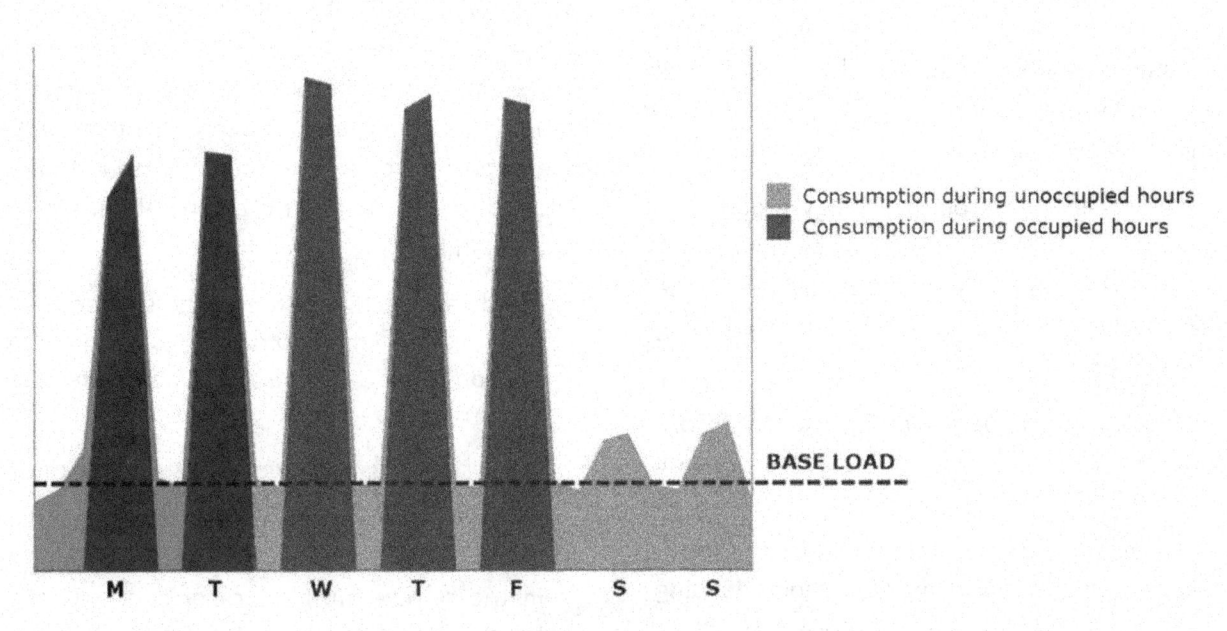

Figure 3.10. Example of electricity data metering result and comparison between occupied and non-occupied hours energy consumption.

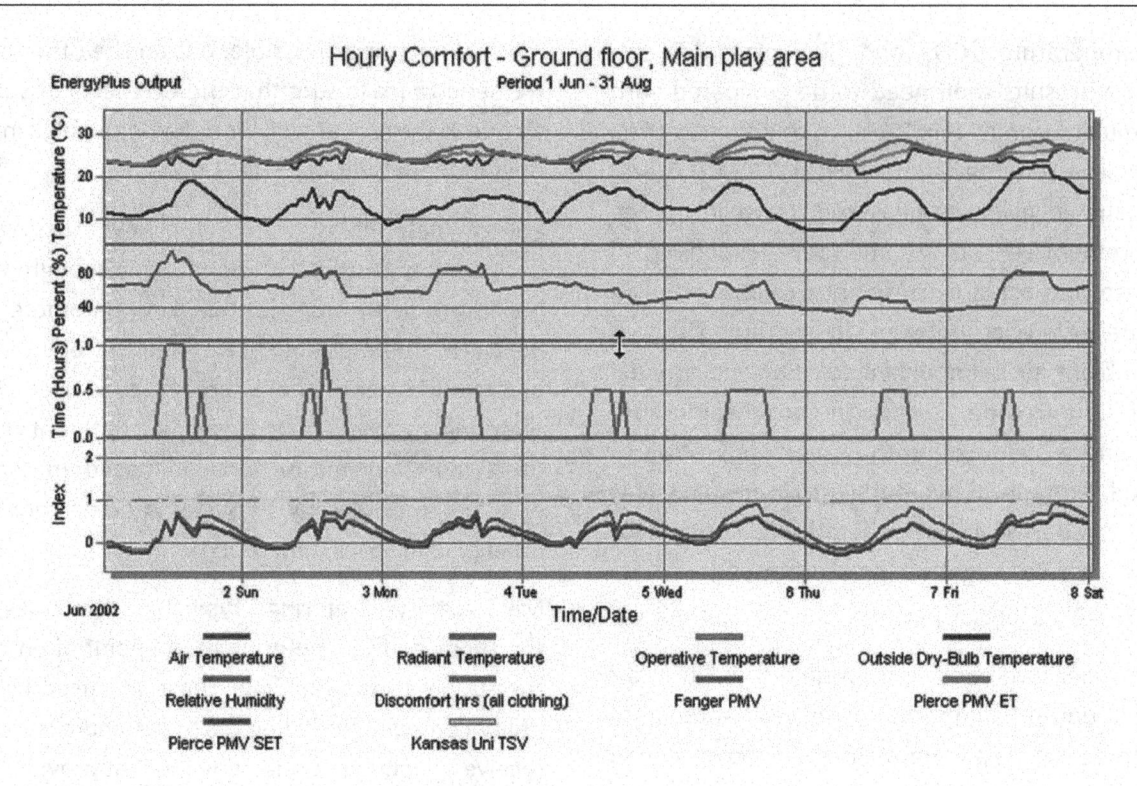

Figure 3.11. Thermal comfort performance simulation results are essential in design phase to evaluate either the performance targets set in OPR are met or not. [27]

Energy load specific sub-metering is installed for different energy loads to measure their consumption (e.g. cooling energy or air handling unit performance). This helps both the building owners and tenants to better manage their energy use in the building and enables them to understand which components are consuming energy while the building is in full use or where the energy is consumed when the building is empty or only partially occupied (Fig. 3.10). Typical HVAC-components to sub-meter are air handling units, chillers and water heaters. Sub-metering should be installed in any individual energy end uses that represent 10% or more of the total annual consumption.

In case a building has on-site energy production, the produced energy needs to be measured. In such According to the USGBC LEED v4, the advanced energy metering must have the following characteristics [67]:

- Meters must be permanently installed, record at intervals of one hour or less, or transmit data to a remote location.

- Electricity meters must record both consumption and demand. Whole building electricity meters should record the power factor, if appropriate.

buildings, energy metering can be done either by measuring both the consumed and the produced energy separately, or by using net-metering.

Energy meters provide useful information on the energy use in a building. Automatic Meter Reading technology provides the facility to read the meter remotely and at frequent intervals. The technology is applicable to both main uti9lity meters and sub-meters. A smart meter allows the utility company to both remotely read and instruct the meter. This additional functionality allows updates of metering software, change of reading frequency and also remote control an agreed portion of a customer's consumption in order to help balance grid electrical generation and demand.

- The data collection system must use a local area network, building automation system, wireless network, or comparable communication infrastructure.

- The system must be capable of storing all meter data for minimum 18 months.

- The data must be remotely accessible. All meters in the system must be capable of reporting hourly, daily, monthly or annual energy use.

Most buildings today measure water only once upon entering the building and do not measure distribution points or end uses. The technology supporting water metering is significantly less developed than that for energy. Potable water metering is typically focused on the quantification of flow volume, not energy content. The measurement technology selected will depend on a number of factors like, current design, budget, accuracy requirements, minimum flow, range of flow and maximum flow. In general, volumetric water metering designs can be broken down into three general operating designs: positive displacement, differential pressure, and velocity including ultrasonic meters.

Automated testing procedures in BMS can help during the functional performance testing phase but also during the on-going commissioning. They utilize advanced programming techniques to reproduce typical commissioning testing both automatically and many systems at once. Automated testing is most effective and economic in systems that are most likely to fail. This includes anything with an actuator such as cooling valves and volume control dampers. Other systems to be included in automated testing are those that consume a lot of energy; units that are hidden or difficult to access and systems that are critical to the operation of the facility. Automating the testing of these units allows the operations to go to the trouble of accessing these units only as needed. The net result is, these units are frequently avoided by the maintenance

Design Review—Commissioning
A review of the design documents to determine compliance with the Owner's Project Requirements, including coordination between systems and assemblies being commissioned, features and access for testing, commissioning and maintenance, and other reviews required by the OPR and Cx Plan.

Design Review—Peer Review
An independent and objective technical review of the design of the project conducted at specified stages of design completion by one or more qualified professionals, for the purpose of enhancing the quality of the design.
(Not part of Cx Process)

Figure 3.12. Difference between commission and peer design review.

and operation staff. If they knew the unit was for in need of repair, they could justify the inconvenience and schedule the repairs.

3.3.6. Design-phase commissioning

Design shall be done based on the Owner's Project Requirements (OPR) and Basis of Design (BoD). Therefore it is very important that OPR includes the performance targets of the building.

Total building performance should be the focus during the entire design and construction project. (Fig. 3.13.) Unfortunately, too often the focus is only on individual component selection and operation. This is mainly due to lack of information of total system operation. Hence it is important to produce right type of operational data during the concept design. Building simulations are required to better understand both the conditions inside the building and the energy use of the building. The typical simulations during design phase are:

- Building orientation and shading;
- Annual energy use of building;
- Daylight levels;
- Lighting simulation;
- Fire and smoke safety simulations;
- Evacuation simulations;
- Thermal comfort simulations.

Such simulations are carried on various specialized software. Energy simulation software can be used both for estimating the annual energy use of a building and the thermal conditions (temperature) inside the building or in various spaces. Computational Fluid Dynamics (CFD) is used to simulate the thermal conditions (temperature and air velocity), indoor air quality, spread of fire and smoke inside the building, and evacuation of occupants in case of emergency. There are also dedicated software for building shading, daylight and lighting simulations to analyse lighting requirements, operating hours of artificial lighting, and optimal orientation of building.

Sequence of Operations (SOO) outlines the engineer's basic intentions on how the mechanical systems operate and are controlled. It is project-specific document that has a direct reflection of the

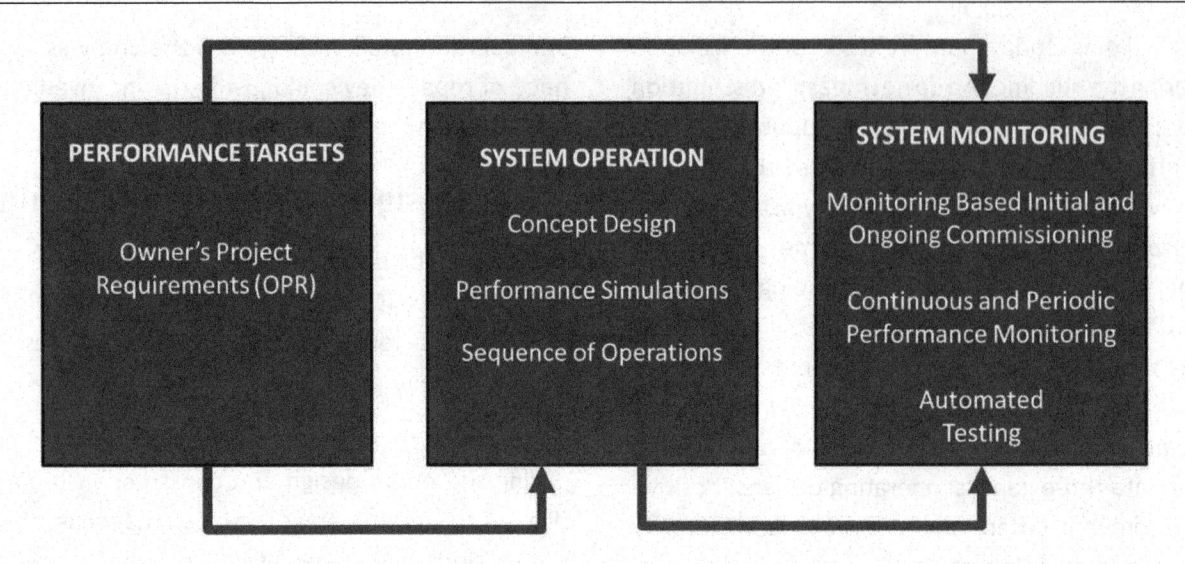

Figure 3.13. The core purpose of commissioning is to ensure that the building performs as the client desired. OPR specifies the targets set by a client. In the design phase the operation parameters are specified more in details (concept design) and the expected performance is verified by simulations. The sequence of operations (SOO) describes in details how each component is supposed to perform in different operation conditions and modes. Monitoring based commissioning would ensure that the defined SOO delivers the expected performance. During the operation the key performance indicators are monitored continuously and in case automated testing procedures are set in BMS, different operational situation could be tested periodically.

mechanical design and it defines all of the required operational points for the project. It needs to be clear, specific, and complete in order to use it as a basis of BMS programming and system commissioning. SOO does not only have the basic operational schemes but also safety limits are included.

The exact set point and limit values in SOO will be finalized during the commissioning as generally the commissioning team will get the best insight of them during the start-up and seasonal commissioning. Also, SOO should not detail the operational characteristics of packaged equipment or describe basic control fundamentals, as it makes it only unnecessarily complicated and too long..

Developing the SOO has four major steps:

1. Create Flow Diagrams;

2. Categorize the Purpose of Each Equipment;

3. Identify Requirements of Codes and OPR;

4. Develop a List of Points.

First step is to create the flow diagram of each system, like ventilation system or cooling system. It should include all the inputs and outputs for the controlled variables. Inputs are those sensor readings that are coming into the building management system (BMS) and outputs are those signals originating from the BMS to the controlled variable.

Next step is to categorize a purpose of each equipment and system and to keep in mind that they may have multiple purposes.

Examples of the purpose of the system are:

- Comfort cooling for occupants;

- Maintaining acceptable temperatures for a process;

- Maintain pressure relationships between two particular spaces;

- Smoke control during a fire.

The designer should also identify any other equipment that is affected by the sequence. As an example, treated fresh air unit needs to be interlocked with the exhaust fans.

When developing the SOO, it is important to ensure that the performance described in the SOO will lead to the expected performance specified by the client in the OPR. Also, all the requirements of various codes, like National Building Code, energy codes, IAQ standards and fire code, need to be fulfilled. It is important to recognize the requirements and exceptions for a particular project location (e.g. earthquakes, extreme weather conditions and traffic). Designer should also confirm whether the owner has any specific operational requirements that are not specified in the OPR and understand how the owner intends to use the equipment or building.

In the end, the designer shall create the list of all points (Tab.3.1.). It identifies all the inputs and outputs that are controlled or monitored in the building, typically by the BMS. The points are then classified as:

• Digital inputs and outputs that are a simple on or off (0 or 1) conditions.

• Analog inputs and outputs that represent a value within a range corresponding to a change in voltage (e.g., 2 to 10 V) or amperage (4 to 20 mA) or a change in air pressure.

• Monitoring capability and alarms are also important to be specified during the SOO development. Storage capability should be defined as well, like:

• Identify how long the data should be retained (e.g., 30, 60, or 90 days).

• The frequency of the trends must also be evaluated (e.g. 30 s or 15 min).

Table 3.1. Examples of points in the Sequence of Operation (SOO) made for the AHU unit. DI = digital input, AI = analog input, DO = digital output and AO = analog output.

Point	Type				Notes
	DI	AI	DO	AO	
Enable/disable			X		
VFD speed command				X	
VFD speed status		X			
Outdoor air temperature		X			
Return air carbon dioxide		X			Alarm above 1200 ppm
Filter differential pressure switch	X				Alarm in 350 Pa

In the end of the design phase, all design documents need to be reviewed in terms of system operation and that the OPR and BoD requirements are met. In particular, the reviews confirm that there are adequate access points, test ports, monitoring capabilities and points, and control features. (Fig. 3.12.) Reviews also verify that energy-efficiency, operation, control sequences, maintenance, training and O&M documentation requirements are consistent with the OPR and BoD. The owner can also extend the depth of the commissioning reviews including the typical peer-review issues like checking and making suggestions for improvements relative to the design concept, bidding, coordination, performance, constructability, sizing calculations, life cycle cost analysis, and code, standards and guideline compliance. Typical items to be reviewed from the system operation point of view are:

• Ventilation system including air handling units:
    • Face velocity in AHU;
    • Specific fan power (SFP);
    • Air filtration;
    • Cooling and dehumidification process;
    • Energy recovery efficiency/effectiveness;
    • Data metering;
    • Duct type, insulation material and leakage test method;
    • Air flow control principles;
    • Location of VCD dampers;
    • Economizer operation;
    • Return air system;
    • Exhaust air system.
• Room air conditioning and air diffusion:
    • Ventilation rates;
    • Pressurization of spaces;
    • Terminal units and air diffusion;
    • Control zones;
    • Data metering;
    • Location of volume control dampers (VCD) and VAV dampers.
• High and low side cooling:
    • Chiller COP;
    • Chiller type (water or air cooled or alternative technology), size and number;
    • Chiller control valves;

- Refrigerants (type, quantity);
- Cooling tower (open/closed loop);
  - Source, management and quality of cooling tower water;
  - Pumping strategy (primary—secondary);
  - Pump size and type;
  - Pipe types and sizing;
  - Pipe insulation material;
  - Differential pressure sensor location;
  - Control valve location, type and size;
  - Isolation valve locations;
  - By-pass pipes and valves.
- Building Management System:
  - The possibility of occupants to influence the room conditions;
  - Room conditions controller types and control parameters;
  - Location of room control panels and sensors;
  - Control principles of air handling units;
  - Air flow management principle;
  - Cooling system controls;
  - Control logic and sequence of operation;
  - Integration of various controls.
- Fire and smoke safety systems:
  - Fire zoning;
  - Type and location of fire dampers;
  - Smoke management;
  - Staircase pressurization;
  - Accessibility of components.

The Cx provider should discuss with the owner the benefits and disadvantages (e.g. impacts to the design schedule and costs) of proposed activities and how many installation reviews are required as the ideal number and timing varies from project to project. Every project should have one design review but in larger ones several at the beginning of each subsequent phase. These reviews should occur at concept phase, during design development phase and during mid-construction documents phase.

The design teams need to provide written response to the design review comments. The Cx provider, design teams and owner shall all understand the situation and agree how to address each comment. This will be done during the commissioning start-up and design review meetings. All decisions should be documented prior to the design team moving into the next phase of design.

Finally, it is important to ensure that all commissioning requirements are incorporated in construction documents. This includes all contractor related commissioning responsibilities:

- Participation in commissioning meetings;
- Provide submittals for commissioning review;
- To use commissioning procedures and checklist during installation and start-up;
- Requirements for documentation and reporting;
- Participation during Functional Performance Testing;
- Measuring instrument calibration requirements;
- User and O&M personnel training requirements;
- To create O&M and System O&M manuals;
- Participation during seasonal commissioning;
- Participation during operational review after 10 months of operation.

3.3.7. Contractors' submittals review

Before components are purchased, the contractor must ensure that all selected components fulfil the design requirements. The contractor shall provide submittal details of all major components to the client, designer and Commissioning Provider, who will check if they fulfil the operational requirements of the project. Cx provider will provide a report to the owner and contractor. Cx provider shall also identify any issues that might otherwise result in rework or change orders. This review does not replace or alter the scope or responsibility of the design team in approving submittals. The following items will be reviewed:

- Product data given by manufacturer has been measured based on the international standards;
- Safety and comfort requirements are fulfilled with selected systems and components;

- Where critical system components are changed by the contractor, the technical performance of the overall system is not affected;

- Selected components and materials have independent, third-party test certificates;

- Products with energy- and eco- labels have been chosen and the environmental impact data is given (if requirement in OPR) ;

- All changes in components, materials and installation is updated to submittal drawings and documents.

The reliability of product performance data is important when comparing different manufacturer's products. There are different kinds of performance data available (Fig. 3.14.). It is important to make sure that when comparing different products, the product data of each option is measured or calculated based on the international standards like ISO, ISHRAE-RAMA, ASHRAE or CEN.

The error in the product data can make the entire system to operate in non-optimum condition that may increase the energy use and reduce indoor environmental conditions.

The most reliable product data is the one that is validated by a third party. There are several international third party validation schemes and certificates available like Eurovent, AHRI or AMCA.

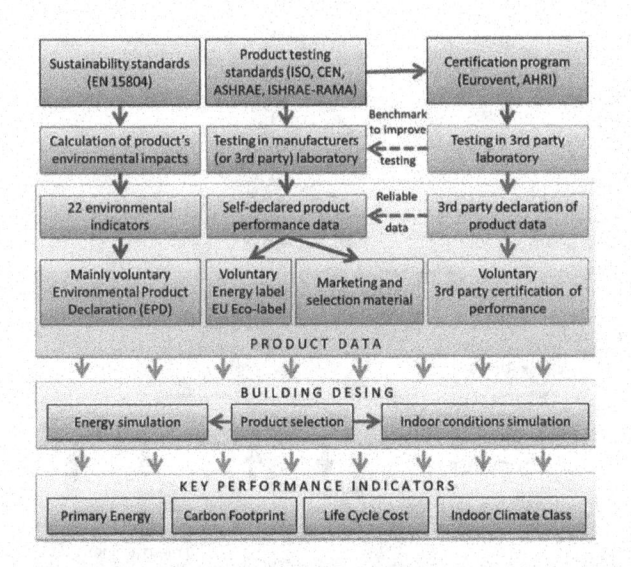

Figure 3.14. Different kind of product data is required during the design when assessing the building performance and when comparing the different manufacturers' data during tendering process.

The performance certificate is then given, based on the performance measurements done in the independent laboratory.

It has lately become more common among manufacturers to give not only the product performance data but also the data about the environmental impact of products. Even it is completely voluntary, it would be important that it is calculated based on the international standards so that it is easy to compare different manufacturers' data.

3.3.8. Installation and inspection of system components

The commissioning team must ensure that the equipment, systems and assemblies are properly installed, integrated, and operating in a manner that meets the Owner's Project Requirements (OPR) and Basis of Design (BoD). (Fig. 3.15.)

Before installation, it is important to ascertain that all components to be installed are as per submittal specification. Therefore each component needs to be checked after arriving at the construction site. The delivery checklists are used to check that the product specification and name plate as-well-as completeness of components and accessories are as per specification. It is also important to visually check that there are no transportation damages. In case the component is not correct or there are transportation damages, the contractor shall inform the supplier immediately to get the replacement delivered to the site as soon as possible. Once the contractor signs the delivery checklist, he approves the completeness of delivery. A Copy of Delivery Checklist shall be given to the supplier.

Installation of pipes, ducts and all system components shall be made based on the standards specified in the design documents. Both the pressure testing of pipework and the leakage testing of ductwork will be made already in the installation phase, before insulating the pipes and ducts. Both of these tests are made as per procedures mentioned in the commissioning plan.

Commissioning provider shall conduct regular visits on site to review equipment, systems and assemblies installation. An objective of the commissioning site visits is to verify proper installation early enough and prevent systematic problems. After each site visit the Cx provider updates the project logbook and sends a report to a client and contractors about any findings during the review. During the next visit, previous defects or other pending items will be reviewed again. In the end of the installation, Cx provider gives an installation review report of completeness of installation and how defects were addressed during the installation.

Commissioning provider shall also review the as-built drawings provided by the contractor especially in terms of system operation.

After the installation of each system or major component is completed, the contractor has to perform pre-commissioning review. Pre-commissioning checklist is required for all operational components as-well-as for ductwork, pipework and BMS as a whole.

For example, the following components need the pre-commissioning review:

- Dedicated outdoor air units, treated fresh air
- Exhaust and supply fans;
- Variable air volume (VAV) dampers;
- Volume control dampers (VCD);
- Fire and smoke dampers;
- Chillers;
- Cooling towers;
- Pumps;
- Unitary products like split AC, cassette units, heat pumps and variable refrigerant flow (VRF) system (indoor and outdoor units);
- Precision air condition units;
- Fan coil units;
- Radiant cooling panels and manifolds;
- Chilled beams;
- Balancing and control valves;
- Measurement sensors;
- Energy (BTU) meters;
- Weather station.

Typical items to be checked are cleanliness, installation, pressure/tightness testing, insulation, connections (air, water and electrical), AHU panel

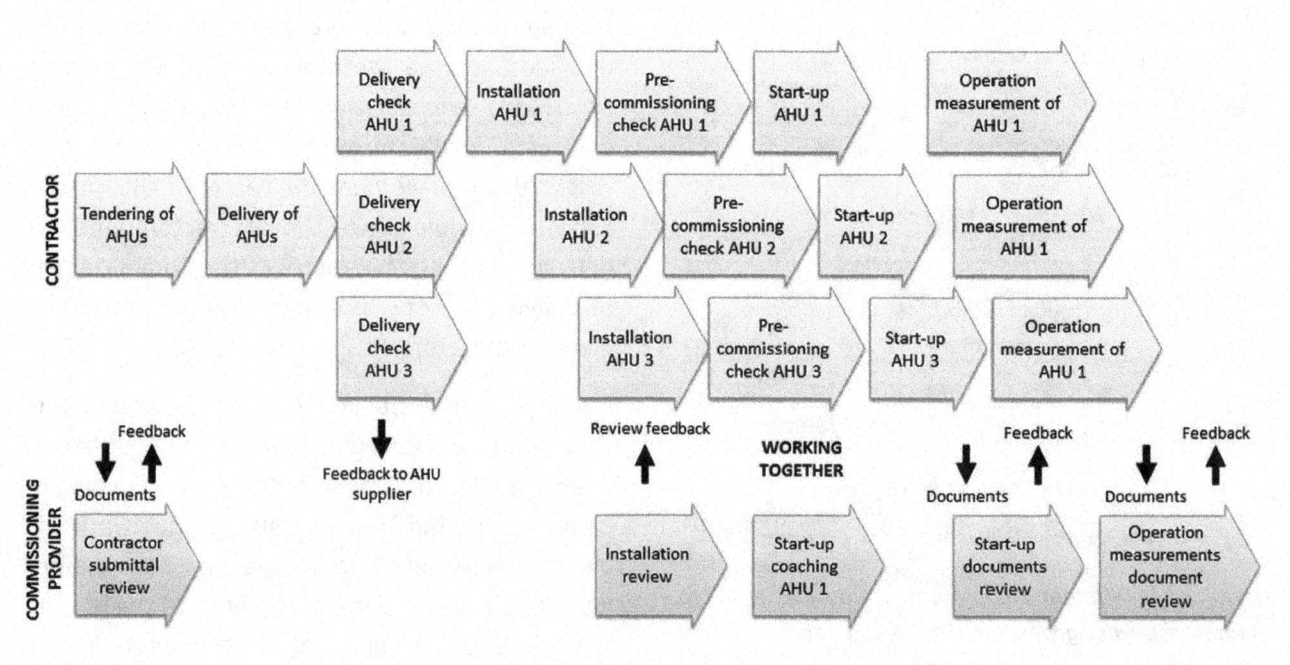

Figure 3.15. Example of AHU tendering, installation and start-up processes and roles of contractor and commissioning provider. During the first AHU start-up, the Cx provider shall assist contractor and make sure that the correct procedures are followed and the contractor has required skills and instruments.

wiring and Variable Frequency Drives (VFD) Installation. If required, make sure that factory-made performance testing results are available during the start-up process. After the pre-commissioning checks, systems are ready to be started.

3.3.9. Component start-up

Component start-up procedures describe how to start the operation of each component. They specify the methods to be used and the checks that are needed. As an example, the following components or systems need to be started:

- Dedicated outdoor air units, treated fresh air units and air handling units;
- Variable Frequency Drivers (VFD);
- Exhaust and supply fans;
- Ductwork balancing;
- Chillers;
- Cooling towers;
- Pumps;
- Pipework balancing;

- Variable Refrigerant Flow (VRF) system;
- Heat pump system;
- Precision air conditioning units;
- Building management system (BMS).

Start-ups and related measurements are carried out by the contractor. Commissioning provider assists the contractor in the beginning of each start-up process and reviews that no systematic errors are made.

Most of the start-ups also require measurements. The contractor must have sufficient measuring devices to carry on tests (Tab. 3.2). Accuracy of each measuring device shall be good enough and the measurement range sufficient to the measured level. It is also important to have each measuring device calibrated less than one year ago. The contractor shall incorporate equipment type, accuracy, measurement range and calibration certificate of each measuring device as part of start-up reports.

All measurement data shall be recorded using the template specified in the commissioning plan. After

Table 3.2. Typical measurements and required devices during HVAC commissioning.

Object	Measurement device	Measurement
Air flow rate	Vane anemometer	Velocity in AHU filter section or grille
	Hot wire anemometer	Velocity inside duct
	Pitot tube and manometer	Pressure difference in duct
	Capture hood & manometer	Air flow rate of grille or diffuser
Air velocity	Hot wire anemometer	Velocity of air
	Vane anemometer	Velocity of air (>0.5 m/s)
Ductwork leakage	Duct leakage tester	Static pressure inside ductwork
Pressure of air	Pressure manometer	Differential pressure in ductwork or e.g. between soaces
	(Barometric) pressure meter	Atmospheric pressure
	Pressure difference data logger	Continuous pressure difference recording
Pressure of water	Hydronic manometer	Differential pressure
	Pressure data logger	Continuous pressure recording

Object	*Measurement device*	*Measurement*
Water flow rate	Hydronic manometer	Pressure across control valve
	Digital flow rate meter	Water flow rate
Temperature	Thermometer	Liquid or air temperature
	Hot wire anemometer	Air temperature (dry bulb)
	Surface temperature meter	Surface temperature
	Infrared temperature meter	Surface temperature
	Thermal imaging camera	Surface temperature
	Sling psychrometer	Wet/dry bulb temperature
	Temperature data logger	Continuous temperature recording
Humidity	Hygrometer	Absolute and relative humidity
	Sling psychrometer	Dew point
	Relative humidity meter	Relative humidity
Indoor air quality	Carbon dioxide meter (NDIR)	CO_2-level of air
	Carbon monoxide meter (NDIR)	CO-level of air
	Particulate meter (gravimetric, light-scattering or beta-attenuation)	Particulate mass concentration and/or particulate count
	Photo-ionization detector (PID)	Various VOCs like TVOC and formaldehyde
	Electro-chemical sensor	Various gases like SO_2, NO_2, O_3
	Andersson active air sampler with petri dish	Total Fungal and Bacterial Count
Refrigeration	Vacuum gauge	Refrigerant system vacuum
	Refrigerant leak detector	Refrigerant leakage
	Refrigeration manifold	Pressure difference in pipe
Filter leak testing	Aerosol photometer and generator	Conditions of the HEPA filter as well as to check installation
Sound	Sound level meter	Sound pressure levels in octave bands, Leq
Fan operation	Tachometer	RPM of rotating objectives
Energy and power	Plug-load meter	Power consumption of device
	Clamp meter	Voltage and current
	Electricity multi-meter	Voltage, current, resistance

the start-up measurements of each equipment or subsystem, the contractor gives all reports for Cx provider, who provides a start-up review report to a client.

(More about pre-commissioning and start-up of individual components in chapter 7).

3.3.10. Functional Performance Testing

The Functional Performance Testing (FPT) is at the core of the commissioning process and it is also the most demanding part of it. It occurs once all the components of the system are installed, energized, programmed, balanced and ready for operation under the partial and full-load conditions. Functional performance testing is the dynamic testing of entire systems (rather than just components) in fully automated mode. Systems are tested under various modes, such as during high or low cooling loads, component failures, unoccupied, varying outside air temperature and moisture, fire alarm, power failure, etc.

Functional performance testing is planned by the commissioning provider and performed by the entire commissioning team including Cx provider, client, contractors, and designers. It may reveal problems with the performance of the commissioned systems and, therefore, may require significant follow-up and coordination among members of the project team to resolve the issue. Equipment operation during the tests is performed by the appropriate contractor or equipment manufacturer, but the responsibility for directing, witnessing, logging and reviewing test data rests with the commissioning authority.

Each component should have a well-defined performance (SOO) as part of the entire HVAC system. Any malfunction can compromise the correct behaviour of the entire system.

The malfunction may be due to:

- Design faults;
- Selection or sizing mistake;
- Manufacturing fault or initial deterioration;
- Installation fault;
- Wrong tuning;
- Control failure;
- Abnormal conditions in use of building.

The functional performance testing (FPT) is devoted to the detection of such possible malfunction and to its diagnosis. The test can be 'active' or 'passive', according to the way of analyzing the component behaviour. An active test assesses performance by analyzing data obtained from systems that are subjected to artificial changes in operational conditions. A passive test will assess performance by analyzing data obtained from systems operating under normal conditions. Active tests are mostly applied during the initial commissioning, i.e. at the end of the building construction phase. Later when the building is in operation, a passive approach is usually preferred in order to preserve health and comfort conditions inside all the building occupancy zones.

Figure 3.16. During the Functional Performance Testing, there is a need to artificially alter the conditions around measurement sensors.

Cx provider creates a plan of functional performance testing of key equipment and systems during the design phase and carries on the performance testing together with the contractor and the designers. The test is planned based on Owner's Project Requirements, but typically it includes tests like:

- Ventilation System Operation (TFAs, AHUs, exhaust fans, VAV boxes and BMS-system) with different weather and occupancy conditions;

- Cooling System Operation (chiller, cooling tower, pumps, control valves and BMS-system) with different weather and occupancy conditions;

- Operation of fire and smoke safety systems (fire dampers, smoke dampers, smoke exhaust fans, staircase pressurization, BMS, supply and exhaust fans);

- Overall Performance of Room Systems in different occupancy/load conditions:

- Monitoring of room conditions: temperature and relative humidity;

- Monitoring of operation of critical components, e.g. VAV-dampers and control valves;

- Indoor air quality measurements like RSPM, CO_2, CO, TVOC and CH_2O.

When a BMS system is part of the project, functional performance testing may use the sensors installed into the system to monitor the performance (Fig. 3.16.) and analyse the results using the system's trend log capabilities. Other option is to use the stand-alone data loggers for data collection.

Overwriting sensor values in BMS system to simulate a condition, such as the outside air temperature reading in a control system, should be avoided when possible. Such testing methods often can only test a part of the system, as the interaction

Examples of Functional Performance Testing

Change in Room Air Temperature

- Ensure that room is cold enough and that the VAV damper 1 in room 1 is in minimum position.

- Measure air flow rates both VAV damper 1 in room 1 and VAV damper 2 in room 2.

- Record a readings in BMS (room air temperatures, VAV damper positions, static pressure in the main duct, fan speed).

- <u>Create a change</u> by increasing the room air temperature in room 1 e.g. by warming up the temperature sensor.

- Record the changes in BMS (room air temperatures, VAV damper positions (max/min and steady-state), static pressure in the main duct (max/min and steady-state), fan speed (max/min and steady-state)).

- Measure the air flow rate in the duct near VAV damper 1 and VAV damper 2.

- Compare the measured and recorded readings to the designed values.

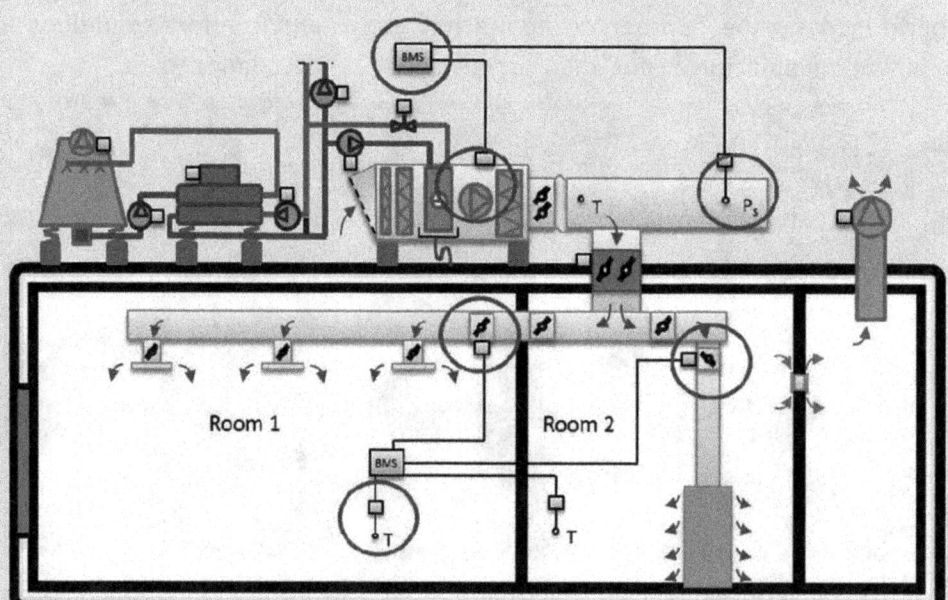

Fire Alarm with Motorized Fire Damper

- Ensure that fire damper is open and both the supply air and exhaust air fans are in operation.

- Simulate the fire alarm in the zone.

- Record the performance of a fire damper, supply fan and exhaust fan in the BMS.

- Review the performance status of all three components in the field.

- Compare the performance status of all three with the Sequence of Operation (SOO) document.

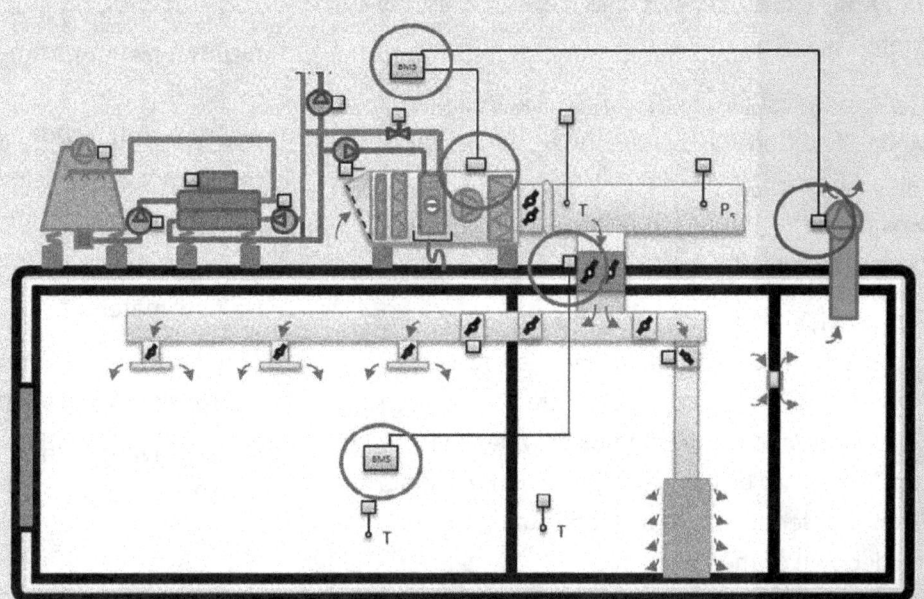

Decrease in Cooling Load

- Ensure that all AHU cooling coil valves are in 100% open position and that the return water temperature is high enough to keep all chillers running with the maximum speed.

- Record all readings in BMS (AHU cooling coil valve positions, inlet and return water temperatures, all pump positions / flow rates, all other valve positions in the chilled water system, differential pressure in the pipework as well as cooling tower fan speed).

- Ensure that all pumps and fans are operating in the building and that the temperature manometers are showing the same water temperatures than in BMS (or measure water temperatures).

- Create a change by closing e.g. 50% of AHU cooling coil valves.

- Record the changes in the BMS (AHU cooling coil valve positions, inlet and return water temperatures, all pump positions / flow rates, all other valve positions in the chilled water system, differential pressure in the pipework (max/min and steady-state) as well as cooling tower fan speed).

- Review the performance status of all components on the field.

- Compare the performance status of all components with the Sequence of Operation document.

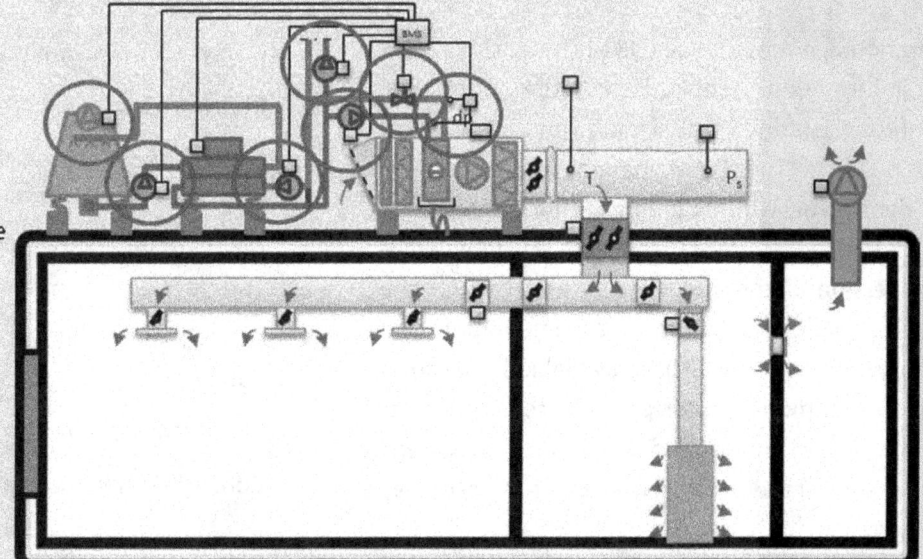

and responses of other systems will be incorrect. Simulating the condition, e.g. the outside air sensor could be heated with a hair drier, should be preferred. Varying the set points to test a sequence is acceptable. Before simulating conditions or overwriting values, sensors and devices must be calibrated.

Multiple identical pieces of non-critical equipment may be functionally tested using a sampling strategy. The sampling strategy is developed by the Cx provider. After three testing failures all remaining units shall be tested.

The Cx provider shall document the deficiency and the adjustments or alterations required to correct the system performance. The contractor corrects the deficiency and notifies the Cx provider that the system is ready to be retested. Corrections of minor deficiencies identified may be made during the tests at the decision of the Cx provider.

If any check or test cannot be completed due to the building structure, required occupancy condition or other deficiency, those functional performance tests may be delayed and be conducted at the same time as the seasonal tests. After the initial occupancy, it is recommended that short-term diagnostic testing be performed, using the building management system to record system operation at least over a two to three week period.

3.3.11. 'O&M Manual', 'System O&M Manual' and 'System Manual for Facilities'

Operation & Maintenance (O&M) Manual contains the information required for the operation, maintenance, decommissioning and demolition of a building. The O&M manual is prepared by the contractor with additional information from the designers, suppliers and the Cx provider. A draft version of the document is provided for the client as part of the handover documentation. The final document is not usually available in full form until several months after practical completion, as commissioning information often needs to include the seasonal readings taken in the fully occupied building.

As per ASHRAE guideline 'The Systems Manual for Facilities' [8], the objectives of the Systems Manual are to:

- Provide the necessary information to the facility operating staff and the maintenance function to understand the design and construction of the facility and how to operate and maintain the building.

- Assemble the facility design, construction, and testing results for building systems in one set of documents.

- Provide a documentation source to be used in training materials.

- Provide documentation for building performance improvement and ongoing commissioning.

- O&M manual shall include the following information:

- Owner's Project Requirements (OPR) document;

- Basis of Design (BoD) document;

- Description of the each HVAC-system that are installed into the building;

- Manufacturer's product data including maintenance instructions, servicing schedules and recommended spare parts listings;

- Maintenance schedules for HVAC Systems and equipment;

- Functional description and control schematic drawings for the BMS;

- As installed drawings including location of key components that require maintenance and access;

- Commissioning data:

- Ductwork balancing tables (air flow rates);

- Pipework balancing tables (water flow rates);

- Chiller and pump measurement data (flow rates, temperatures, pressures);

- Air flow and pressure measurements of AHUs and operating frequency;

- Control and balancing valves settings (pressure and flow);

- Volume control dampers settings (position).

- Commissioning plans and reports:

- Commissioning plan;

- Commissioning Design Review report;

- Commissioning Submittals Review report;

- Testing and start-up reports, evaluation checklists, and testing checklists;

- Cx Progress Reports;

- Issues and Resolution Logs;

- Item Resolution Plan for open items.

- Guarantees, warranties and certificates.

- Particular requirements for demolition, decommissioning and disposal.

The contractor shall also develop a System O&M manual (Fig. 3.17.) that gives existing and future operating staff the information needed to understand the system operation in different seasons and operating conditions. It should be understandable to people unfamiliar with the project. The system O&M manual generally focuses on operation rather than the maintenance of the equipment, particularly the interactions between different components. It should include the following:

- Operation of systems:

- Operating plan;

- Facility and equipment operating schedules;

- Set points, ranges, and limitations;

- Sequence of Operations (SOO);

- Emergency procedures.

- System single-line diagrams.

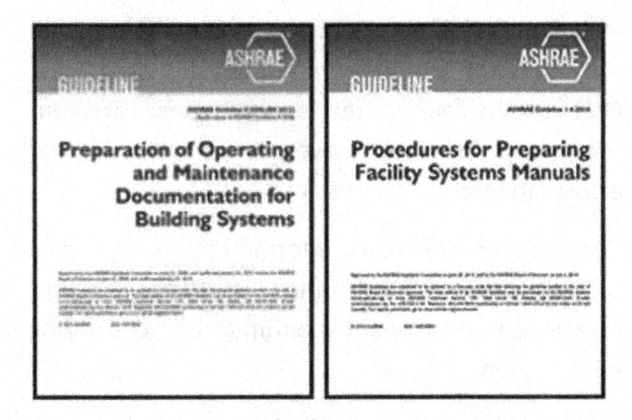

Figure 3.17. There are two ASHRAE guidelines focusing on the O&M documentation: Guideline 4-2008 - Preparation of Operating and Maintenance Documentation for Building Systems and Guideline 1.4-2014 - Procedures for Preparing Facility Systems Manuals. [9, 11]

- As-built sequences of operations, control drawings, and original set-points.

- Operating instructions for integrated building systems:

- Ventilation system operation in different seasons/weather conditions;

- Cooling system in different seasons/weather conditions;

- Fire and smoke safety systems.

- Recommended schedule of maintenance requirements and frequency, if not already included in the project O&M manuals.

- Recommended schedule for retesting of commissioned systems with blank test forms from the original commissioning plan.

- Recommended schedule for calibrating sensors and actuators.

- Indoor air quality measurement schedules if not fully integrated to BMS.

- Ongoing commissioning activities and schedules.

- The System Manual for Facilities includes also the maintenance and training plans in addition to the above:

- Maintenance plans, procedures, checklists and records:

- Maintenance schedules;

- Utility measurement and reporting;

- Ongoing commissioning operational and maintenance record keeping;

- Janitorial and cleaning plans and procedures.

- Training:

- Training plans, materials and videos;

- Training records;

- Operator's ongoing documentation of modifications and adjustments to the facility systems and assemblies.

3.3.12. Training of O&M Personnel and Building Users

The key objective of the owner's operating and maintenance personnel training is to transfer the knowledge and skills required to operate the building. This includes an understanding of the OPR and BoD as well as training on the purpose and use of O&M and System O&M Manuals.

The Commissioning provider reviews the contractor's submittals of the training content, materials, and instructor qualifications. The owner may designate to have the Cx provider participate in training sessions and/or use other methods to confirm that the training was delivered effectively. The training may include the following topics:

- General purpose of system;

- Use of O&M manuals;

- Review of control drawings and schematics;

- Normal day-to-day operations in the building;

- Start-up, normal operation, shutdown, unoccupied operations, seasonal changeover, manual operation, control setup and programming troubleshooting and alarms;

- Interactions with other systems;

- Adjustments and optimizing methods for energy conservation;

- Health and safety issues;

- Special maintenance and replacement sources;

- Occupant interaction issues;

- System response to different operating conditions;

- Emergency situations.

It is also important to provide orientation and training to occupants relative to building systems and the assemblies they interact with. This is especially important when new technologies are used in a building.

It is highly recommended that all trainings be recorded as this allows for future reference of the material and training of new personnel as they join the team.

3.3.13. Seasonal commissioning activities

Building operational parameters are dependent on the weather conditions. As the initial commissioning is done during the few weeks, it is very unlikely that different weather conditions occur during that time. Therefore, not all aspects of the building

commissioning can be carried out during the normal contract period. Seasonal commissioning hence, is required to recognize those items of the system operation that are either weather or occupancy related.

The importance to have the operational parameters of ventilation and cooling systems for summer, winter and monsoon cannot be diminished. Also all systems need to be fine-tuned to real operational conditions. It may be the case that the building is not operating with full occupancy load or that the design conditions were either over or under estimated.

During the seasonal commissioning the following items need to be addressed:

- The building and its services operate correctly, safely and efficiently;

- Equipment is tested and adjusted continuously, balanced and fine-tuned to achieve the specified performance whilst optimizing flow rates, temperature, etc., for each season;

- Automatic controls have been set-up and tested;

- Any design faults and over/under specifications are highlighted;

- Systems are clean and properly maintained;

- The system settings and performance test results have been recorded and accepted as satisfactory.

Additional information may be used for this, including feedback from tenants about what they think of temperatures in specific areas, etc. One way to collect user feedback is to carry out the user satisfaction survey of perceived indoor environmental conditions.

Seasonal commissioning information must be used to make necessary adjustments to system settings to ensure that the plant is optimized for use during each season.

The O& M and System O&M -Manuals should be updated with seasonal commissioning data. O&M personnel are informed how to keep the Systems Manual up to date as changes occur throughout the life of the building.

3.3.14. Operational review after 10 months of operation

It is important to assure that the building's performance is maintained during the first year of the operation, particularly before the warranty period expires. Therefore operation of systems and components are critically reviewed by the commissioning team 10 months after occupancy to identify any items that must be readjusted or repaired or replaced under the warranty. Inconsistencies between design performance and actual performance and/or an analysis of any complaints received from users may indicate a need for minor system modifications.

Before conducting the operational review meeting, the commissioning provider have to make sure, that seasonal commissioning has been carried out and the O&M and system O&M manuals are updated accordingly.

Conducting both an O&M personnel interview and a User Satisfaction Survey before a review meeting would be advantageous. The O&M personnel should be surveyed to evaluate building equipment controls and performance. User survey is presented to the occupants 6 to 12 months after move-in to confirm that a satisfactory Indoor Environmental Quality (IEQ) has been achieved for a substantial majority of the occupants. Survey shall focus on how the users have perceived different IEQ elements: thermal comfort, indoor air quality, acoustic conditions, lighting conditions and cleanliness.

A building's energy performance needs to be benchmarked as well. After the initial commissioning, energy consumption and other variables such as time of day, weather, hours of operation and occupancy should be monitored for 6-12 months. Correlations should then be developed between whole building energy use and the variables and in larger buildings also between major equipment and variables. Actual energy consumption shall be then compared with designed energy consumption and other benchmark data available. In case there is a big variation compared to benchmark data, the reasons need to be found out and corrective actions taken. Reasons for both big and small consumption need to be studied. This can be a sign of some technical failure, non-correct set values, etc. Final conclusion of the building's energy consumption can only be done after the second year of operation, when all operational errors have fixed.

Any other key performance indicator set in the OPR needs to be reviewed during the operational review of the building. Therefore it may be that some measurements are needed prior to review, like indoor air quality measurements. Typically also, issues like room air temperature or relative humidity has some set and limited values. This kind of data can be collected from the building management system. Besides, other post-occupancy measurements, like air flow rates in air handling units can be considered before operational review.

walk-through in a building can also bring more information of failures in installation, but also issues related to indoor environmental quality like odours or noise. It may also give an indication of additional commissioning requirements (e.g. high velocity in occupied area or stable air may be signs of poorly balanced ventilation).

Ongoing training includes refresher training of existing personnel, training of new personnel and training of all personnel on newly installed equipment or revised operating procedures.

trouble-shooting and lessons-learned workshop between O&M personnel, designers, CxA and contractors needs to be organized. A workshop should be conducted to discuss and document any problems O&M personnel have had when operating the building and to discuss project successes and identify opportunities for improvements for future projects.

The commissioning provider invites all stakeholders to conduct a Building Operations Review on-site 10 months after substantial completion, typically near the end of the warranty periods. The Operational Review includes the review of the following data:

- Results of the User and O&M personnel surveys;

- Indoor environmental quality measurement results;

- Walk-through audit report;

- Energy consumption analysis;

- Data of other key performance indicators (set in OPR);

- Report of alarms and/or other data indicating system failures;

- Data of user and O&M personnel trainings;

- Work orders related to commissioned systems;

- Trend logs and equipment operation;

- O&M and System O&M manuals.

Issues identified during the review are documented along with a proposed solution and identification is made of the responsible party for correction. Issues under warranty of the original construction contract are provided to the contractor to be taken care. Any additional commissioning is also identified and determined to either contractors or O&M personnel. If there is a need for additional training, that is also discussed and agreed. The Commissioning Report is updated to reflect the Operational Review and other changes or additions that have occurred during the building operation.

After the Operational review all systems need to be optimized based on the opportunities identified during the review. Fine-tuning the system performance includes items like optimizing schedules, sequences, and set-points in addition to other no-cost/low-cost changes. The CxA may assist in implementation of the changes.

The Operational review is the beginning of an Ongoing Commissioning Program. It includes the repeating of the functional testing portion of the commissioning process on a periodic basis, or ongoing monitoring and trending with associated automatic or manual diagnostics, or a combination of these methods. (Further details about ongoing commissioning in chapter 5).

3.4. Commissioning cost

The total building commissioning costs for the services of the Commissioning Authority can range from 0.5% to 1.5% of total construction costs [58]. The National Association of State Facilities Administrators (NASFA) recommends budgeting 1.25 to 2.25% of the total construction costs for total building Commissioning Authority services. U.S. General Administration Services' (GSA) commissioning practice is expected to cost approximately 0.5% of the construction budget for federal buildings and border stations. More complex projects such as courthouses could run 0.8-1% of the construction budget, and even more complex facilities such as laboratories can exceed can exceed 1% (Fig. 3.18.) [32].

There are many factors influencing commissioning costs. The scope of work is one of the most important factors. Currently in India the commissioning cost is much below the cost levels in US, both absolutely and relatively. One of the reasons for low fee is that the Commissioning Authority's scope of work in India is currently limited to installation and start-up reviews.

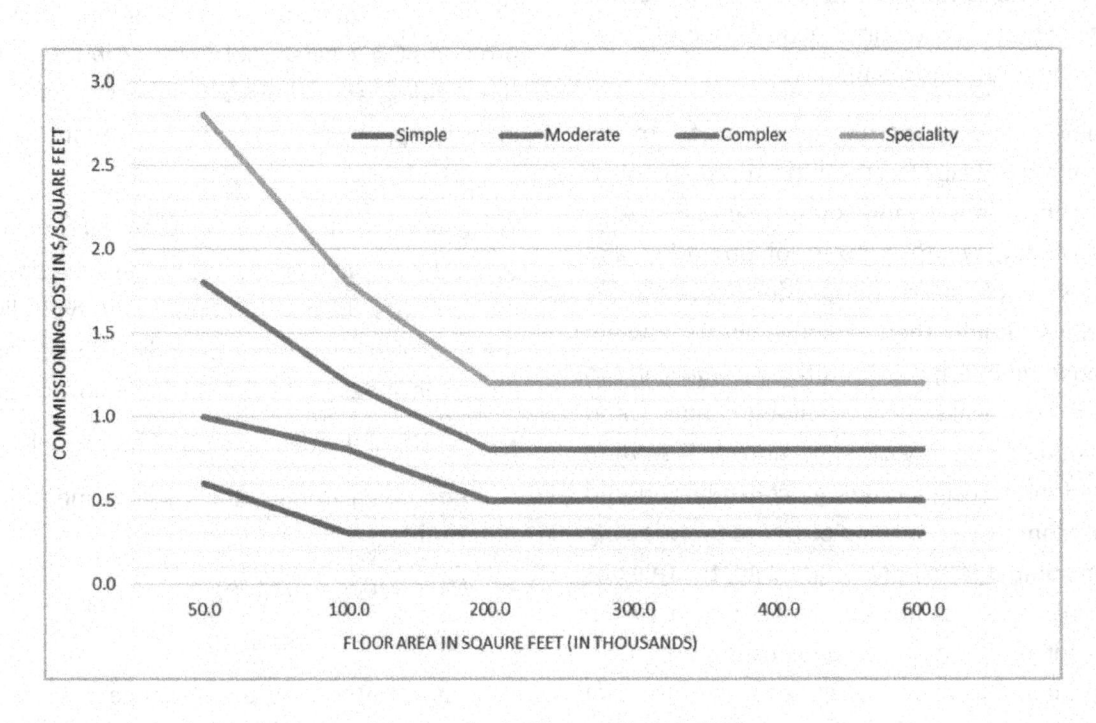

Figure 3.18. Commissioning Authority services costs by facility type in U.S. [32].

4. RETRO-COMMISSIONING

Retro-commissioning (RCx) is the commissioning process in existing buildings. It seeks to improve the system performance together with indoor environmental quality. Depending on the age of the building, retro-commissioning can often resolve problems that occurred during design or construction, or address problems that have developed during the building's use. In all, retro-commissioning fine tunes and/or upgrades a building's operations and maintenance (O&M) procedures to enhance overall building performance.

Typical reasons for retro-commissioning are:

- Improve occupant comfort;

- Identify operations and maintenance and energy efficiency improvements;

- Train the building operators on how to help improvements persist;

- Review and enhance building documentation.

4.1. Planning of retro-commissioning

Retro-commissioning will help to keep the building operating at optimal levels. The timing of a retro-commissioning effort will vary depending on the timing of changes in the facility's use, the quality and schedule of preventive maintenance activities, and the frequency of operational problems. Factors that indicate the need for retro-commissioning include:

- An unjustified increase in energy use;

- An increase in the number of complaints from occupants about IEQ;

- An increase or relatively high level in night-time energy use.

Retro-commissioning needs to be planned well. Typically, it consists of five different phases:

- Planning of retro-commissioning activities;

- Investigation of existing documentation and data collection on site;

- Design of required repair, retrofit and system adjustment activities;

- Actual repair, retrofit and adjustment activities;

- Validation of result by commissioning authority.

4.2. Investigation of existing documentation

Commissioning work always requires the drawings and documentation of current design and component selections. However in many cases the current documentation is not up-to-date or may not be found at all. This is why in retro-commissioning, sufficient time needs to be reserved for finding and updating the existing documentation. The following documents should be available:

- HVAC drawings (contractor's submittal drawings);

- Initial Basis of Design documents (HVAC, electrical, plumbing);

- Electrical drawings and other design documents;

- List of all equipment used in projects;

- O&M manual or technical data sheets of each equipment;

- Ductwork and pipework balancing records prepared after latest balancing (e.g. after ductwork cleaning);

- Initial start-up measurement reports of AHUs, chillers, fans, pumps, etc.

Out of these documents, the HVAC drawings and equipment details are the most important documents.

4.3. Understanding of current system operation and IEQ

Before actual retro-commissioning activities can be started, the current HVAC-system operation needs to be understood. Required data can be found partially from BMS system, if that exists. Typically BMS has the following information:

- Operating schedules and start-up/shut-off criteria of each equipment:

- Air handling units;

- Treated fresh air units & DOAS units;

- Exhaust fans;
- Chillers;
- Pump
- Room conditions (air temperatures, relative humidity and CO2);
- Energy use of HVAC-systems;
- Load variation trends of fans, pumps, chillers, etc;
- Water consumption.

Walk-through audit is an important part of retro-commissioning. It is recommended to have the audit before the detailed system measurements. During the audit the focus shall be on:

- Energy conservation opportunities (ECO);
- Water saving opportunities;
- Unusual odours and their sources;
- Moisture damages and mould growth;
- Condition of all equipment, piping, ducting, wiring and related auxiliaries.

Additional air quality tests are often required in various locations. The following tests shall be considered: Respirable Suspended Particulate Matter (RSPM), gases (e.g. CO_2, CO, TVOC, CH_2O, SO_2, NO_2, O_3) as well as total fungal and bacterial count. The actual performance of each equipment and system needs to be measured and audited (Fig. 4.1. and Fig. 4.2.).

The following measurements are typically required:

- AHU/TFA/DOAS:
- Air flow rates;
- Pressure loss across each component (each filter, cooling coil, energy recovery wheel, etc.);
- Fan speed and pressure across the fan;
- Air temperatures: ambient, return, before cooling coil, after cooling coil, after fan section;
- Cooling water temperatures (entering/leaving) and water flow rate;
- Input power (motor).
- Exhaust fan:
- Air flow rate;
- Air temperature;

- Fan speed;
- Pressure difference across the fan;
- Input power (motor).
- Chiller:
- Evaporator and condenser water flow rates;
- Evaporator and condenser chilled water temperatures (entering/leaving);
- Refrigerant load and quality;
- Suction and discharge pressure of refrigerant;
- Compressor oil quality and level;
- Cleanliness and fill of refrigerant;
- Input power.
- Cooling tower:
- Water flow rate;
- Chilled water temperatures (entering/leaving);
- Fan speed;
- Water quality;
- Ambient air temperature.
- Pumps:
- Water flow rate;
- Pressure difference (head);
- Input power.
- Ventilation system:
- Air flow rates in main ducts;
- Air flow rate in each diffuser;
- Supply air temperatures;
- Exhaust air volumes, especially in critical environments like laboratories, operating theatres, etc.
- Room pressurization:
- Spaces compared to other spaces;
- Spaces compared to the ambient air;
- Spaces compared to the return air shaft.
- Hydraulic system (e.g. radiant cooling, floor cooling, chilled beams, fan coil system):
- Water flow rate in main pipes (control valves);
- Water temperatures (entering/leaving).

- Fire and smoke safety systems:

- Operation of staircase pressurization fans and dampers;

- Review of fire damper operation;

- Smoke exhaust system operation review.

- If the detailed energy audit is done at the same time, the following issues need to be carried out:

- Preparation of process flow charts;

- Preparation of all service utilities system diagrams;

- Collection of annual energy bills & tariff data and preparation of energy consumption patterns;

- Analyse energy supply sources (e.g. electricity from the grid or self-generation);

- Analyse potential for fuel substitution, process modifications, and the use of cogeneration systems (combined heat and power generation).

Based on the audit results, there are two major parameters to be calculated for each equipment:

- Efficiency = input/output (e.g. chiller kW/TR);

- Loading = actual output/designed output.

The efficiency and loading gives a good overall picture of current systems operation.

One important item to be considered is the life cycle and current condition of equipment, pipes, ducts and wires.

4.4. Design and planning of required improvements in system operation

Retro-commissioning is executed based on the owner's expectations of the future performance of a building. In case the OPR was not documented during the initial construction, it needs to be created before retro-commissioning starts.

It is also possible that requirements have increased since the initial commissioning phase in terms of indoor air quality or energy use of the system. If there is no separate Basis of Design (BoD) document, the OPR needs to be detailed enough in terms of component performance to enable the evaluation of the results of the retro-commissioning process in the end of the project. Based on the data collected in the building, the required improvements are planned.

Typically there are different kinds of improvement required in terms of implementation time and cost. Some improvements may require major refurbishment (e.g. adding a new outdoor air ventilation system), and others can be done as part of annual repair, retrofit and maintenance schedule. Typical activities to improve HVAC-system operation, IEQ, energy efficiency and safety are:

- AHU/TFA/DOAS adjustments (e.g. air flow rates, cooling coil temperature, filtration, fan speed);

- Repair and cleaning of AHU/TFA/DOAS (e.g. damaged sound attenuation material surfaces, dirty or rusted condensation collection tray);

- Repair of fans (e.g. broken belt or motor, operation of fan impeller, bearings, lubrication);

- Replacement of fans (e.g. new EC fans);

- Chiller repairs or adjustments (e.g. chilled water temperature levels, condensing water temperatures, condenser pressure control, refrigeration loads, compressor oil pressure and quality);

- Cleaning of evaporator coil from dust;

- Repair of refrigeration pipework;

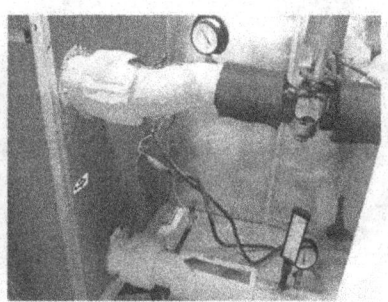

Figure 4.1. The performance, condition and cleanliness of all equipment need to be measured and reviewed.

- Replace pressure and temperature gauges if missing or damaged;
- Cooling tower repair and/or cleaning;
- Replacement or repair of pumps;
- Ductwork repairs (leakage, insulation & hanging brackets);
- Ductwork cleaning;
- Ductwork balancing;
- Hydraulic system balancing;
- Replacement of control valves;
- Repair pipework insulation, hangers and brackets;
- Improvements in instrumentation (replacements or new measurement points);
- Retrofitting sensors to monitor new parameters that were not considered during the initial design;
- Replace damaged wires, safety switches, thermostats, controllers, contactors, protection devices, etc.;
- Repairing anti-vibration mountings;

- Re-program operation schedules in BMS;
- Repair of any flooding, moisture or mould damage (either in equipment, pipework, ductwork or in structures).

Cost estimations of each activity and corresponding payback analysis are made during the planning phase. There are also non-monetary benefits like improvement in users' health and comfort.

4.5. Planning and execution of commissioning Activities

Commissioning process and planning follow mainly the same steps as during the initial commissioning. Both individual component adjustments and comprehensive functional performance testing is required. Before they are carried out, any repairs or retrofits required need to be completed and pre-commissioning checks need to be carried out.

System components are adjusted using the same procedures as during the initial start-up. It is also important to update both O&M and Systems O&M Manuals and provide necessary training to O&M personnel.

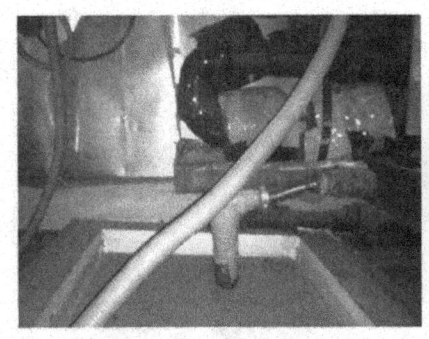

Figure 4.2. Sometimes the equipment items are at the end of their lifetime or the desired performance cannot be reached and therefore need to be replaced before the system is adjusted to the new operation mode.

Typical Energy Conservation Opportunities in European Buildings

The 'iSERV cmb' project [44] was about showing the practical operation and benefits of an automatic monitoring and feedback system to Heating, Ventilation and Air-Conditioning (HVAC) systems in EU Member States. The project established the continuous monitoring and benchmarking of HVAC processes to provide energy saving benefits and to produce benchmarks of energy consumption by HVAC systems. It was also directed to helping specific system managers and owners to identify and implement energy conservation opportunities (ECO). The project listed some typical ECOs, that were sorted, first, by payback and then by capital cost.

- **Priority 1 - low payback and capital cost:**
  - Shut off A/C equipment when not needed;
  - Shut off auxiliaries when not required;
  - Shut chiller plant off when not required;
- **Priority 2 - low / medium payback and capital cost:**
  - Modify controls in order to sequence cooling and heating;
  - Reduce power consumption of auxiliary equipment;
  - Split the load among various chillers;
  - Apply variable flow rate fan control;
  - Install variable volume pumping;
- **Priority 3 - medium payback and low capital cost:**
  - Improve central chiller / refrigeration control;
  - Consider conversion to VAV;
  - Perform night time ventilation (in cold and temperate climate);
- **Priority 4 - medium / high payback and high capital investment:**
  - Reduce compressor power or fit a smaller compressor;
- **Priority 5 - high payback and high capital investment:**
  - Replace ducts when leaking;
  - Eliminate air leaks (AHU, packaged systems);
- **Priority other:**
  - Re-pipe chillers or compressors in series or parallel to optimize circuiting;
  - Consider cool storage applications;
  - Reduce motor size (fan power) when oversized;
  - Sequence heating and cooling;
  - Operate chillers or compressors in series or parallel;
  - Check (reversible) chiller stand-by losses;
  - Clean or replace filters regularly;
  - Switch off circulation pumps when not required.

Similar retrofit opportunities shall be considered in all buildings during the retro-commissioning. In India, the ranking could be slightly different as the labour costs are cheaper than in Europe.

http://www.iservcmb.info

5. ONGOING COMMISSIONING

Ongoing commissioning (OCx) is a continuation of the initial commissioning processes during the occupancy and O&M phase. OCx verifies that a project continues to meet current and evolving OPR. Ongoing commissioning process activities occur throughout the life of the facility; some of these activities will be continuous in nature of their implementation, while others will be either scheduled or unscheduled. In order to achieve objectives and identify smart solutions, facility managers must implement better technologies along with a new management culture built on best practices. Ongoing commissioning is a building management process that focuses on measured building operation data and its continuous improvement.

Implementing ongoing commissioning practices within operations is one of the key steps in creating a smart building. Ongoing commissioning contributes to a building's operation and maintenance. Unfortunately, even today we have too much reactive maintenance in buildings, i.e. simply responding to day-to-day issues. Preventive maintenance includes maintenance or changing filters, checking belts and lubricating bearings.

Ongoing commissioning is a core part of proactive maintenance. It includes activities that look for issues before they become a bigger problem like reduction in indoor environmental quality or hidden energy or water consuming systems. Predictive maintenance includes processes that measure critical system parameters and attempts to predict the point at which a system will fail. Good ongoing commissioning should also include some predictive elements. Ongoing commissioning is an eight-step process that must be a part of the annual building management cycle (Fig.5.1). It starts after the first year of operation, which is still considered to be part of initial commissioning phase.

It includes the measurement and follow-up of various KPIs, management of benchmark data, comparing results with benchmark data, updating a property strategy, developing an annual action plan, preparing a budget for the required resources and implementing required actions.

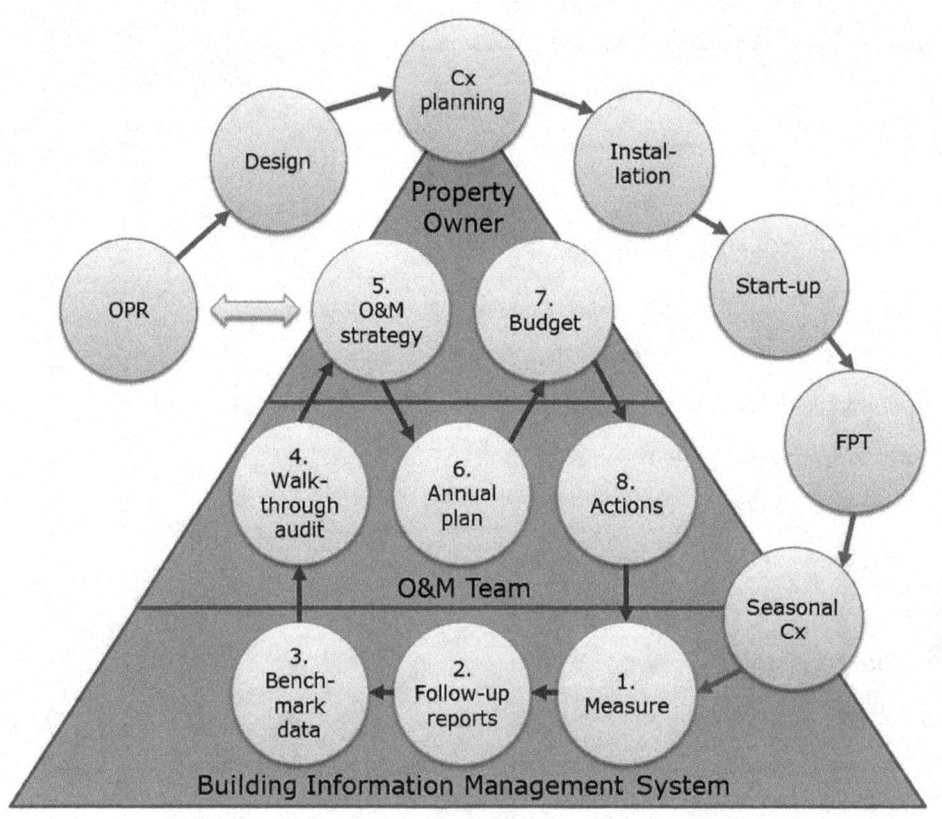

Figure 5.1. Ongoing commissioning process involves different layers of building management.

Ongoing commissioning plan should include issues like:

- Overview of project;
- Long and short-term goals: OPR or O&M strategy including Key Performance Indicators (KPI);
- Roles & responsibilities of different stakeholders;
- List of systems/equipment/parameters to be included and their performance criteria (part of Systems O&M manual);
- Detailed commissioning procedures and reporting requirements for all systems and equipment (part of Systems O&M manual);
- An annual maintenance schedule of all activities;
- BMS alarm plan with reporting and response;
- Automated test forms and procedures;
- List of parameters to be systematically followed up in the BMS and requirements for trend data collection;
- Data collection plan of periodic measures: why, what, how, when, who;
- User feedback system and how/who to respond;
- Annual/monthly/weekly/daily follow-up plan: why, what, how, when, who;
- Training schedule of O&M personnel;
- Janitorial and cleaning plans and procedures.

5.1. Data monitoring and measurements in building

Efficient data monitoring and measurements enable Monitoring-Based ongoing commissioning of the building. Data measurement can be divided into several different categories:

- Continuous monitoring of performance parameters as part of BMS;
- Automated testing of systems as part of BMS;
- Periodic measurements and data collection;
- Random measurements when required.

Continuous monitoring is typically planned during the initial design and measurements are integrated into Building Management System. Typically these are issues that are easy to measure online or are critical to a building's operation. Continuous data monitoring typically includes:

- Energy use (total and sub-metering);
- Water use (total and sub-metering);
- Room conditions: temperature, RH and CO_2, sometimes also VOC and RSPM.
- Periodic measurements and data collection complements the continuous data metering. Typically these are related to:
- User satisfaction, e.g. user satisfaction index based on annual user survey.
- Indoor air quality, e.g. RSPM, CO, O_3, TVOC, CH_2O, SO_2, NO_2, and total microbial count;
- Refrigerant leakage checks;
- Acoustic environment, e.g. sound levels in various parts of building;
- Water quality, e.g. potable water, chilled water and water in cooling towers;
- Energy costs, e.g. electricity and gas;
- Water cost.

In case there is no continuous data monitoring in the building, basic data regarding room conditions, energy and water use need to be collected periodically. There should also be an online system for users to give feedback from system operation and any other problems related to indoor environmental quality.

A typical period depends on the type of data, but most of the data needs to be collected 1-3 times per year.

Random measurements are typically related to some unexpected incident, like water leakage, heavy storm, flooding or fire. These can be moisture measurements, material samples or system audits. There are also some IAQ parameters like radon or lighting levels that need to be measured once only, if no major changes are made in the building.

Automated testing is recommended for systems that are most likely to fail. This includes anything with an actuator such as cooling valves and volume control dampers. Other systems to be included in automated testing are those that consume a lot of energy, units that are hidden or difficult to access and systems that are critical to the safety or operation of the facility (e.g. fire dampers). Automated testing procedures are typically designed as part of data monitoring during the design phase.

Automated testing can be considered as an example with the following systems:

- VAV-system: performance of motorized damper actuators and reheating coil valves;

- Radiant cooling and chilled beams: performance of control valves;

- AHU/TFA/DOAS: performance of motorized dampers (e.g. economizer), energy recovery wheel motor and cooling coil valves;

- Fire dampers: actuator operation;

- Smoke exhaust system: operation of smoke extract damper actuators and fans.

5.2. Follow-up of the building performance

Different stakeholders require data continuously from the system operation. Therefore, there is a need to create automatic reports to various stakeholders daily, weekly, monthly and annually.

Typically, most of the alarms need to be handled either immediately or on the same day. Therefore, the BMS system shall send the immediate alarm to the responsible person who takes actions to resolve the issue. However, it is also good to have monthly

alarm reports available for O&M management to be reviewed, that all problems have been solved. Typical alarms include:

- Fire or smoke alarm;

- User complaint;

- Too high or too low air temperature, RH or CO_2 level in BMS;

- Too high or low water temperature in BMS;

- Shut-off of fan or pump in abnormal time (against operation schedule);

- Sudden pressure loss in water pipe (possible water leakage);

- High filter pressure loss.

System shall create reports both from execution of planned and unplanned maintenance activities. It is important to follow-up. All activities of an annual action plan must get done on time. Equally important is to analyse the reasons behind unplanned maintenance activities in order to manage cost and resources better and improve ongoing commissioning activities.

Executives shall also get their monthly reports regarding energy and water use as well as IEQ conditions. These should be short reports

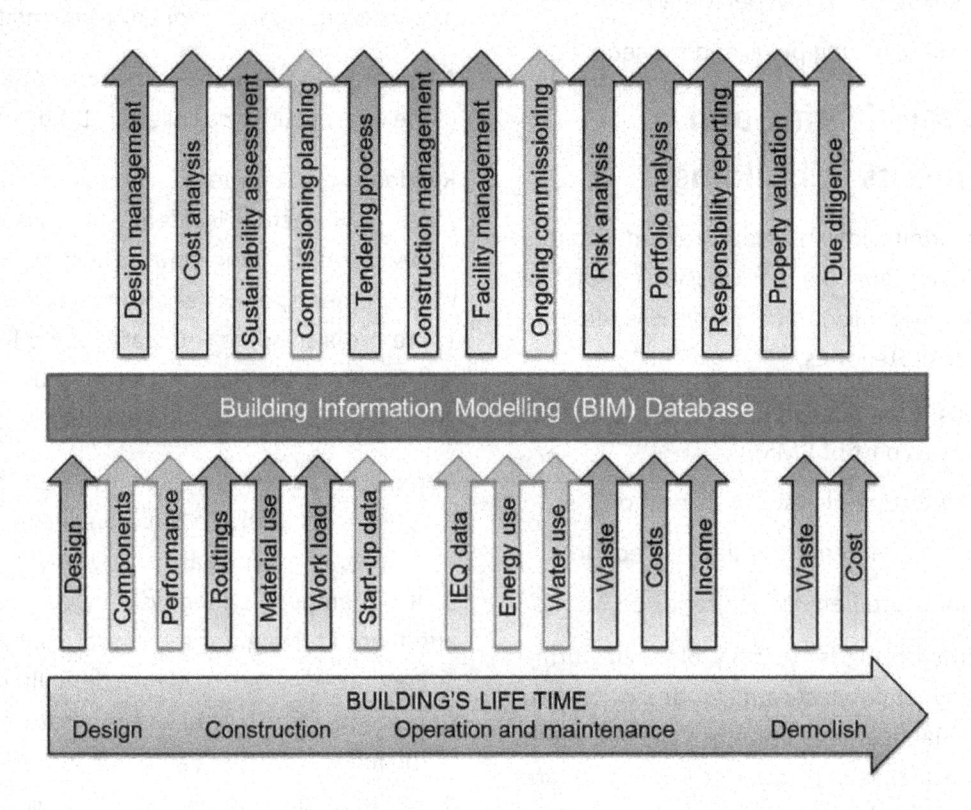

Figure 5.2. A building's performance data should be collected throughout the lifetime of the building i.e. product manufacturing, design, construction, commissioning, facility management and demolishing. [69]

highlighting the current status of some key indicators compared to benchmark data. Same data complemented with sub-metering data must be shared with O&M management.

Building users should also be aware of key performance data of building in order to adapt their behaviour when using the building. This can be implemented in different ways, e.g. by providing online data service in the company's' intranet or having screens at the entrance of the lobby to communicate the current performance.

Building information modelling (BIM) system is used to store the designed, commissioned and operational data over the life cycle of a building. (Fig. 5.2.) During the design phase the values of performance indicators are collected from different performance analysis e.g. energy and indoor conditions simulations. Also, during the construction and commissioning, the BIM model will be automatically updated with any changes and collected performance data. Different performance and sustainability analysis and reports are easy, fast and cost-efficient to make.

If all performance related data is saved and continuously updated in the BIM model throughout the lifetime of the building. It also provides information required for Commissioning Planning and Ongoing Commissioning and any data generated during the commissioning processes shall be stored into the BIM system.

5.3. Benchmark data for comparison

Benchmarking is the process of comparing your system performance to something similar. 'Something similar' might be internal, for e.g. performance at the same time in the previous year. It might be external, for e.g. performance compared to similar facilities elsewhere. Comparing the data without a proper benchmark data makes it difficult to evaluate whether the performance is good or bad. (Fig. 5.3.)

Benchmarking works best when it is done consistently over a period of time. In a recent study, EPA in US found that buildings that were benchmarked consistently reduced energy use by an average of 2.4 per cent per year, for a total savings of 7 per cent. And, buildings that started out as poor performers saved even more. [31]

A different kind of benchmark data is required in each case depending what is the metric that is benchmarked and what kind of data is available. Benchmark data can be found in various sources:

- Design data of a building;
- Building Performance Simulation Data;
- Previous hours/days/months/years data of the building;
- Owner's other buildings that are used in a similar way, e.g. office building 10/5 or hotel building 24/7;
- Common benchmark databases, e.g. BEE or IGBC for energy efficiency;

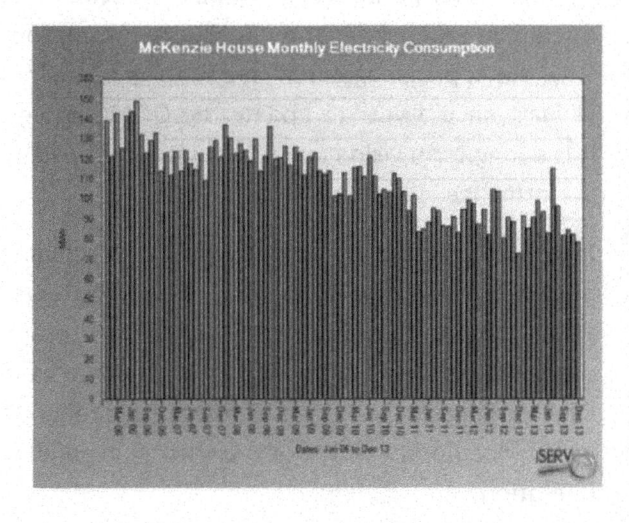 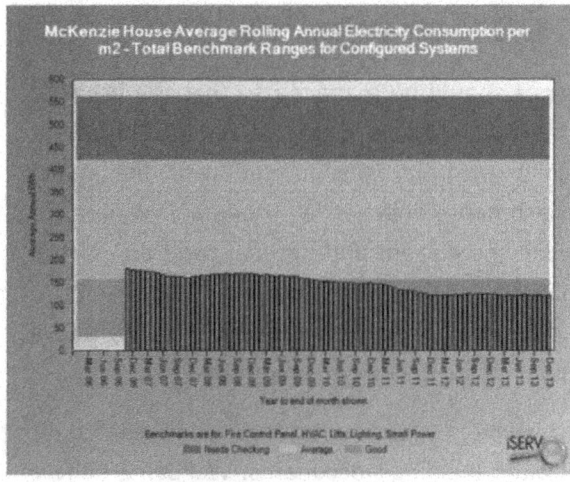

Figure 5.3. Benchmark data to make correct conclusions during performance analysis. [34]

- Legislation, e.g. Energy Conservation Building Code;

- Standards, e.g. ISHRAE 10001 for IEQ.

Also the benchmark data needs to be updated continuously and be saved, e.g. to BIM database.

5.4. Annual walk-through audit, follow-up meeting and reporting

Walk-through audit needs to be done at least once a year in a building. During the audit the main focus shall be on issues reducing indoor environmental quality and increasing energy or water use. Similar issues need to be reviewed or measured as during retro-commissioning (see chapter 4.3). Any Energy Conservation Opportunity (ECO), water saving opportunity and possibility to improve IEQ needs to be recorded.

After walk-through audit, all stakeholders (e.g. representatives from owner, O&M team, property manager, service vendors, users (HR and health care team) shall meet and discuss findings and potential reasons for them. In the same meeting it is good to review user feedback, alarm records and both planned and unplanned maintenance activities and key performance indicators.

After the meeting the facility management team needs to prepare a report of KPIs and other findings and recommendations with cost and time estimates. Required improvement activities shall be prioritized based on their urgency (immediate/short term/ medium term/long term) as-well-as capital cost and payback time (low/medium/high).

5.5. O&M strategy

Buildings should be managed so that the value of the building remains high and no refurbishment or wellbeing debt is generated over the years. Refurbishment debt is a monetary difference between the current and optimal condition of the building and its technical systems. It is the amount of money that needs to be invested to upgrade a building and its systems to an optimal level.

Refurbishment debt generates wellbeing debt among the building's occupants. Wellbeing debt is

the temporary or permanent reduction in human health and wellbeing due to poor IEQ. Systematic maintenance of a building that focuses also on indoor environmental quality reduces the time when people are exposed to poor IEQ. This in turn minimizes the wellbeing debt (Fig. 5.4.). Unfortunately the symptoms often take time to disappear, even if the indoor environment is improved and the refurbishment debt is diminished.

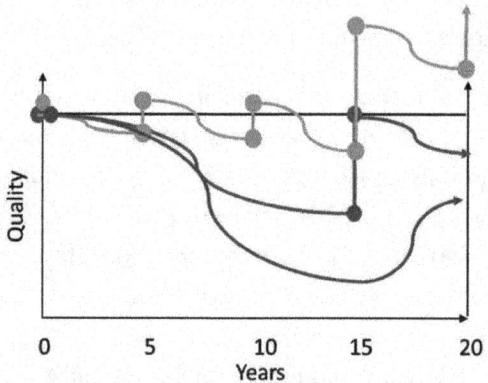

Figure 5.4. Wellbeing debt grows slowly and after the refurbishment, it disappears slowly, sometimes leaving permanenthealth damages to the occupants.

A building's value is dependent on a user's willingness to pay a higher rent, reduced operation and maintenance cost and risks related to investment in the future. The higher the refurbishment debt, the higher are the risks related to investment, thereby reducing the value of the building. Higher the wellbeing debt, more likely it is that rent level is low or users leave the building. Therefore, there is a clear linkage in a building value, between its operation and maintenance. These are the reasons why the top management needs to get involved at least once a year to update the O&M strategy. It is a long term vision and sets targets for building performance.

It should focus on various aspects, including building value, expected annual income, IEQ target values, energy, water use and cost targets and other environmental impacts. The basis of the O&M strategy is the Owner's Project Requirements (OPR) document and KPIs defined for building operation.

O&M strategy should include as an example the following elements:

- Updated KPIs and target values;
- Long and short term targets for a building's operation and maintenance;
- Refurbishment plan (long term);
- Repair and retrofit plan (medium and short term plan);
- Annual maintenance plan;
- Long and short term financial and manpower budgeting.
- The O&M strategy can also be common for several buildings in case their usage profile is similar.

5.6. Annual O&M, repair & retrofit planning, budgeting and implementation

After the top management has finalized the O&M strategy, the annual plan for each building needs to be created. This includes elements, like:

- Detailed maintenance schedule, e.g. filter changes and equipment/systems audits;
- Detailed training plan for new and existing O&M personnel;
- Detailed annual testing and adjustment plan;
- Detailed IEQ and water quality measurement schedules;
- Detailed plans for annual repairs;
- Detailed plans for retrofits and refurbishment carried out in coming year.

These plans shall include all the required activities, new equipment and other materials as well as the manpower plan and cost budgets.

After the annual O&M plan is ready, the budgeting of costs and the required manpower need to be part of the company's annual budgeting process. During the implementation of each plan, it is important to ensure that personnel fully understand what they are supposed to do and why. It is also recommended to have the continuous auditing in place to establish the quality of all O&M activities.

Date	Who identified issue	Issue	Initial recommendation how to solve issue	Probable benefits to owner if solved	When to be resolved	Who to resolve (person & organisation)	Who to supervise (person & organisation)	Steps taken towards solving issue	Status at the moment (open / closed)	Closed date and signature	Confirmed to be resolved

Figure 6.1. Example of a commissioning log book.

6. COMMISSIONING MANAGEMENT

6.1. Commissioning team and role of commissioning provider

The commissioning team consists of various stakeholders during the construction process. (Fig. 6.2.) Each project shall have a commissioning provider (or authority or agent), who is in charge of managing and planning all commissioning activities and is a leader of the commissioning team. The Commissioning team consists of Cx provider, client representative, architect, MEP designers and contractors.

There may also be other stakeholders like green building consultant and manufacturers of critical components, whose participation is useful. The client nominates the commissioning provider for each project. It can be the client's, designer's or contractor's representative or an independent third party company or person.

The best approach for commissioning provider depends on the complexity, type and size of the project. In larger or technically more challenging projects, where a high quality of commissioning is required, the third party commissioning authority is recommended. In case the project seeks the LEED certification, the independent third party Cx provider is mandatory. Cx provider must be totally objective when leading the commissioning process and must maintain an unbiased approach to problem solving and conflict resolution.

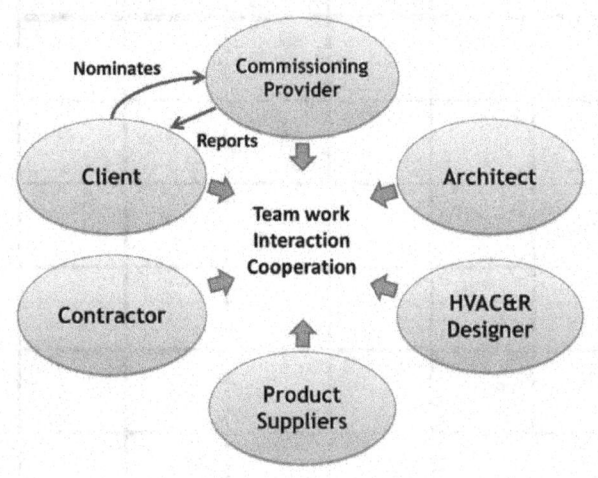

Figure 6.2. Commissioning organization is adapted according to the complexity and quality expectations of a given project.

The following skills are required from the commissioning provider:

- Maintains an unbiased approach to problem solving and conflict resolution;
- Has knowledge of HVAC systems, covering design, common control strategies, installation, operations and maintenance;
- Is experienced in controls for HVAC systems, including familiarity with current technology;
- Has practical background in field construction;
- Has demonstrated ability to organize many specific activities into a coherent commissioning plan;
- Has good communications skills, both written and verbal;
- Is proficient in documentation;
- Is experienced in working with multidisciplinary teams;
- Is familiar with testing and balancing.

6.2. Third party commissioning

Third party commissioning provider is nominated to coach the rest of the team to implement all commissioning activities well. Therefore, the work consists of planning, training, auditing and testing activities. It is important to understand that the third party commissioning provider is not carrying out commissioning activities, but assisting the team to do so. The commissioning provider typically has the following tasks during the commissioning:

- Reviews owner's project requirements and basis of design;
- Develops the commissioning plans: initial, seasonal and ongoing commissioning;
- Develops Functional Performance Testing (FPT) plan;
- Ensures that all commissioning activities are transferred to construction documents (e.g. tender documents);
- Carries out commissioning process review meetings;

- Reviews pre-design and design documents (in case the client has not nominated a separate per reviewer consultant);

- Reviews or creates pre-commissioning and equipment start-up procedures and checklists;

- Ensures that contractors have required skills and tools to perform all commissioning activities and provides training if required;

- Reviews the contractor's submittals (critical equipment and submittal drawings);

- Reviews installation of critical components in terms of safety and system operation;

- Reviews start-up of equipment and systems like pipework tightness, ductwork leakage or AHU operation;

- Carries out the functional performance tests;

- Reviews both 'O&M manual' and 'system O&M manual';

- Ensures that O&M personnel is trained to understand the system operation;

- Prepares the final commissioning report;

- Carries out the system operation review after 10 months of use.

6.3. Commissioning meetings and management of log book

Typically, four meetings are required during the initial commissioning process and one after ten months of operation.

Start-up meeting is held after the commissioning provider is nominated to coordinate the time-schedules and roles of commissioning provider, project manager, LEED-consultant, designers, architects and contractors. In this meeting the owner's project requirements and key performance indicators are reviewed as well as targets regarding LEED-certification.

Second commissioning meeting is held after Cx provider has reviewed the design documents and finalized the commissioning and functional performance test plans. These plans and all start-up procedures and checklists will be reviewed and discussed during the meeting. In case the contractors are not aware of all procedures and measurement activities, separate training will be organized. All pending issues on site will be reviewed.

Third commissioning meeting is held after the contractor has finished the pre-commissioning checks and the Cx provider has reviewed the submittals, pressure & leakage tests and installation. The final coordination (content and time schedule) between contractors and Cx provider is done in order to carry out start-up procedures and operation testing of equipment as well as functional performance testing. All pending issues on site will be reviewed.

Fourth commissioning meeting is held after the Cx provider together with the client, designers and contractors have completed the functional performance testing. Also the O&M manuals and training are reviewed. In this meeting the FPT results are compared with the owner's project requirements and key performance indicators. All pending issues will be reviewed and the action plan to complete them is agreed upon. After the fourth commissioning meeting, the Cx provider prepares the final commissioning report.

Operation review meeting is carried out after 10 months' operation review. Cx provider together with the client and the designer have reviewed the system operation and done a walk-through audit in the building. During the meeting the key performance indicators are reviewed. Based on the seasonal commissioning activities by the O&M personnel (and contractor if agreed so), the system O&M manual is updated. After the operational review, the client can prepare the final checklists for the contractors before 1-year warranty ends.

During the entire commissioning process, the commissioning provider keeps the logbook (Fig. 6.1.) of all findings and pending issues. The Logbook should include the data like date, who identified the issue, what is the issue, initial recommendation to solve the issue, probable benefits to owner if solved, steps taken towards resolving the issue, when to be resolved, status at the moment (open/closed), who to resolve (person & organization), who to supervise (person & organization), closed date and signature, and date and signature of who confirmed to be resolved.

Table 6.1. Summary of Commissioning activities and role of Commissioning Authority in each phase.

	Commissioning Process Activities	Client	Designer	Contractor	Cx Provider
Pre-design	• Prepare Owners' project requirements (OPR) including KPIs; • Prepare Basis of Design (BoD); • Prepare commissioning plans.	• OPR	• BoD		• Cx management • Cx plans • Review of OPR and BoD
Design	• Development of pre-commissioning, start-up and operation review checklists, data collection sheets and procedures; • Functional Performance Test (FPT) plan; • Planning of data metering; • Ensure that all commissioning requirements are included in the contractors' tender documents.		• Operational simulations • Planning of data metering • SOO • (Peer design review)		• Development of pre-commissioning, start-up and operation review checklists, data collection sheets and procedures • FPT plan • Cx design review
Installation	• Contractor's' submittals: equipment details and building drawings; • Pipe- and ductwork pressure/leakage testing; • Product delivery and pre-commissioning reviews - completeness and quality of installation.		• Contractor submittals review	• Su bm itta ls documents • Product delivery reviews • Pressure/leakage testing • Pre-commissioning reviews	• Contractor submitta ls review • Review of pressure and leakage testing • Installation reviews
Start-up	• Balancing and adjustment of systems; • Start-up and testing of all operational components; • BMS inputs testing; • Functional Performance Testing (FPT); • System O&M manual; • Training of O&M team & users.	• FPT	• FPT	• Balancing of systems • Start-up activities • Operational measurements • FPT • System O&M manual • O&M training	• Assistance in start-up, testing and balancing • Random testing of operations • Review of start-up and testing documents • FPT • Final Cx report
1st year operation	• Seasonal commissioning; • Post-occupancy measurements (user satisfaction, IAQ, etc.); • Operation review after 10 months of operation. • O&M training and trouble shooting session	• SeasonalCx • Operational data • Update of O&M manual • O&M training and trouble shooting session			• Operational review

7. TYPICAL PRE-COMMISSIOINING AND START-UP PROCEDURES

7.1. Ductwork installation, cleanliness and insulation

Air is distributed in a building by means of connected individual duct sections that make up the ductwork. The configuration of a ductwork is comparable to a tree with branches connected to the terminal unit. The fan is located at the air handling unit (AHU). In reality, ductwork forms a double-tree because the fan is in the middle of the supply and the return/outside air parts of the system. Tees, crosses or transitions usually connect duct sections.

When installing ductwork:

- Ductwork should be kept as straight as possible. Any turns, bends, extensions, reductions or S-loops will cause additional pressure loss and may end up to reduced airflow;
- No building cavity should be used as ductwork, e.g. panning joist or stud cavities;
- All ducts and mechanical equipment shall be installed within the conditioned building envelope;
- No ductwork should be installed in exterior walls.

The following criteria must be considered for the duct assembly and its elements [54]:

- Dimensional stability (shape deformation and strength);
- Containment of the air being conveyed (leakage control);
- Vibration (fatigue and appearance);
- Noise (generation, transmission, or attenuation).
- Exposure (to damage, weather, temperature extremes, flexure cycle, wind, corrosive atmospheres, biological contamination, flow interruption or reversal, underground or other encasement conditions, combustion, or other in-service conditions).
- Support (alignment and position retention).
- Seismic restraint.
- Thermal conductivity (heat gain or loss and condensation control).

Duct velocity is a critical issue in terms of pressure loss and sound generation—the lower the velocity is, lower is the pressure loss and sound level. Therefore, the smallest branch ducts should be designed for duct velocity of less than 5 m/s (1,000 fpm) that equals the pressure loss of 1-5 Pa/m. Larger ducts in the floors are designed to velocities up to 10 m/s (2,000 fpm) and in the biggest main ducts up to 20 m/s (4,000 fpm).

Duct leakage refers to the fact that the air inside a supply air duct is under a positive pressure. It will leak out of the Pittsburg or Snap lock seams, from the slip, drive or TDC connector joints or out of wall penetrations from damper rods, screws used in hanging ducts or any wall penetrations. (Fig. 7.1)

Leaky supply ducts don't deliver the air where it is end up to reduce airflow; needed. When air leaks out of a ductwork some. No building cavities shall be used as ductwork, areas at the end of the run may be short of airflow e.g. paining joist or stud cavities; because the air has leaked from the system before it. All ducts and mechanical equipment shall be reaches its intended location. This lack of air can install within the conditioned building cause undercooling in the summer or inability to envelope; heat in the winter or poor air quality. Leaky return No ductwork should be installed in exterior walls duct add load. Leaky return ducts can also pull in air from uncontrolled spaces, causing humidity problems, condensation and contamination.

Duct insulation and sealing, especially insulated supply ducts delivering conditioned air within a building, save energy. The intent is to keep mechanically warmed or cooled air as close to a constant, desired temperature as possible and prevent the conditioned air from the escaping the duct system while it is being moved to spaces, where it is needed. If reduced heat transfer through insulated ducts is accounted for in the heating, ventilation and air conditioning (HVAC) load calculations, it may even be possible to reduce the size of HVAC equipment. It has been estimated that 10-30% of heated or cooled air is lost through ductwork.

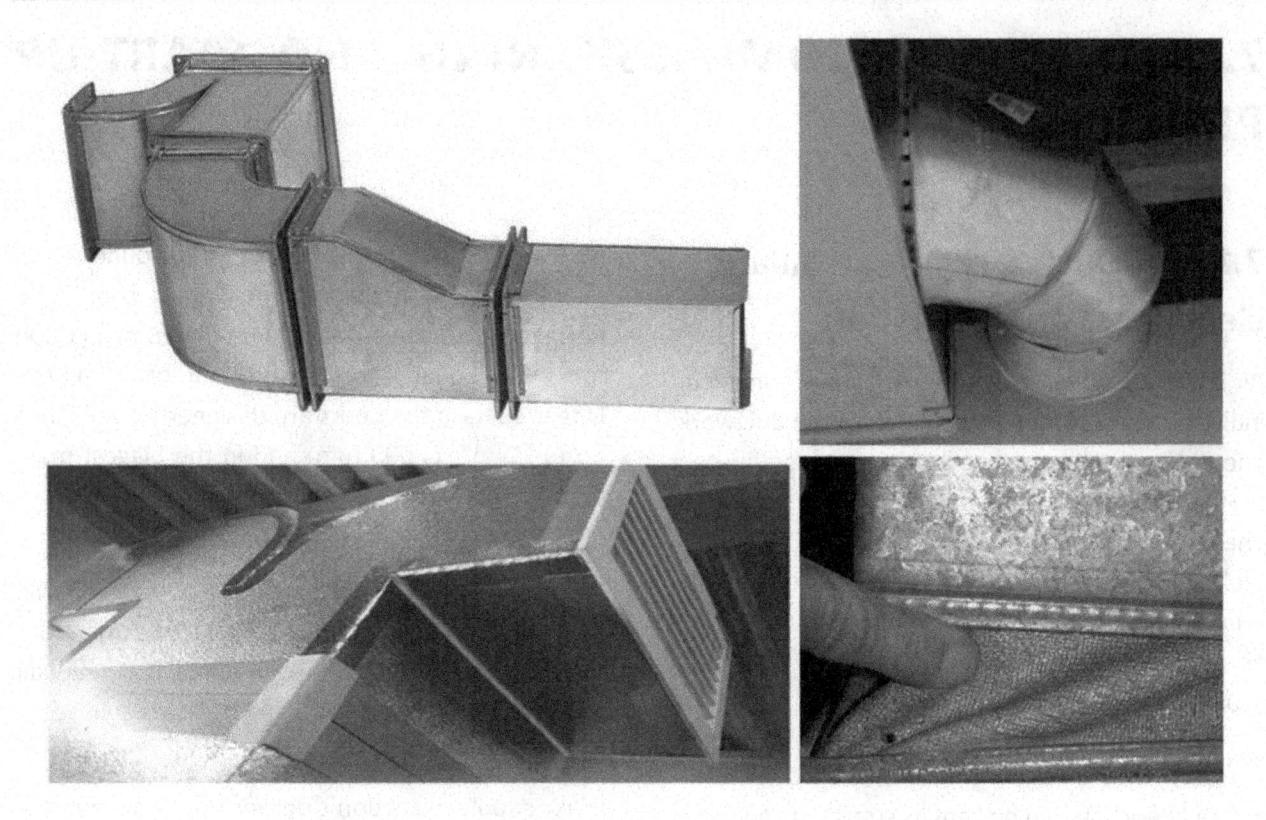

Figure 7.1. Typical leakage points in the ductwork.

Ductwork insulation shall be done only after ductwork leakage testing. Often the air inside the duct is colder than air dew-point around the duct.

Duct insulation also prevents condensation from forming, as the surface temperature of the insulation is higher than the surface temperature of the duct. Condensation will not occur, provided that the insulation surface is above the dew point temperature of the air and that the insulation incorporates some form of water vapour barrier that prevents water vapour from passing through the insulation and condense on the duct surface.

Therefore, the following items need to be checked in terms of insulation:

- Sufficiency of insulation material: thermal conductivity, water vapour resistance, surface emissivity and insulation protection from ambient conditions like solar, temperature and rain;

- Thickness of insulation is as per design documents;

- There are no damages in the insulation material surface;

- Insulation material is well connected to the pipe surface so that there are no air pockets between insulation and duct.

The following are the locations where leakage is most likely to occur:

- Transverse joints: duct to duct, branch, tap;

- Joining of two edges in the direction of airflow;

- Penetration-rod, wire, tubing, etc.

Prior to application of insulation and painting, all installed ductwork, including exhaust, smoke extraction, air conditioning, ventilation, etc., shall be tested for air leakage. The method of air leakage test shall follow 'Low pressure duct construction standards' and 'High pressure duct construction standards' issued by SMACNA (Sheet Metal and Air Conditioning Contractors National Association) of USA. [61]

Ductwork pressure testing is a method to determine how well ducts are constructed to prevent air leakage. Duct leakage testing pressurizes a closed section of ductwork to a known pressure. level The amount of leakage at a specified pressure is calculated by measuring the amount of air that is measured as it is blown into the closed duct system.

SMACNA has defined duct leakage rates as Class 24, 12 or 6. These numbers simply mean that at 1 inch (250 Pa) of test pressure ducts can be expected to leak 24, 12 or 6 CFM per 100 square foot of duct surface. [61]

Maximum leakage is calculated using the following formula:

$$F = C_L * P^{0.65}$$

Where

- F = Max Leakage (cfm/100 ft2)
- CL = Leakage Class: CFM Leakage per 100 ft2 @ 1 inch of H2O (250 Pa) (from table 4.)
- P = Pressure (inch of H2O)

Table 7.1. SMACNA duct leakage classes C_L for rectangular and round metal ducts. [62]

Sealing class	C	B	A
Applicable static pressure construction class	2 inWG (500 Pa)	3 inWG (750 Pa)	4 in WGor higher (1,000 Pa)
Leakage C_L: rectangular metal	24 CFM (11.31/s)	12 CFM (5.71/s)	6 CFM (2.81/s)
Leakage C_L: round metal	12CFM (5.71/s)	6CFM (2.81/s)	3CFM (1.4 1/s)

The mere presence of duct sealing material on a joint seam is no guarantee that the ductwork is properly air sealed. Once the installation contractor has leak tested several sections of ductwork for leaks and has adequate documentation that the duct sealing procedures being used actually keep the leakage within specified ranges, the frequency of sealing can be reduced be reduced.

SMACNA has specified the duct sealing requirements for duct systems. If ductwork is not sealed, leakage rates of 25% on 2 inch (50 mm) plus ductwork can be expected.

SMACNA seal classes of A, B and C (Tab. 7.1. and Fig. 7.3.) define the degree of sealing requirements to be completed. For Seal Class C, all Transverse joints (Slip, Drive & TDC) connections need to be sealed; for Seal Class B, all longitudinal seams (Pittsburg and Snap lock) and all joints as above need to be sealed. Seal Class A is the most stringent it means that all wall penetration (damper rods, screws, and duct accessories) must be sealed in addition to all seams and joints as defined above. Failure to properly specify sealing or to actually seal ducts in the field are the primary reasons for ductwork systems to leak. There is no need to verify leakage control by field testing when adequate methods of assembly and sealing are used.

During testing and commissioning, the following procedure should be adopted:

- Ductwork systems shall be cleaned by removing the dust using the supply air fan, or by using a robot as recommended by the ductwork system cleaning specialist if employed (Fig. 7.2.);
- Temporary filter media shall be used where building work is still in progress during testing and commissioning, and replaced with clean filters for final measurements of flow rates;
- Computer room plants, in particular where under floor air distribution systems are used, should not be run before the rooms and floor voids have been properly cleaned;

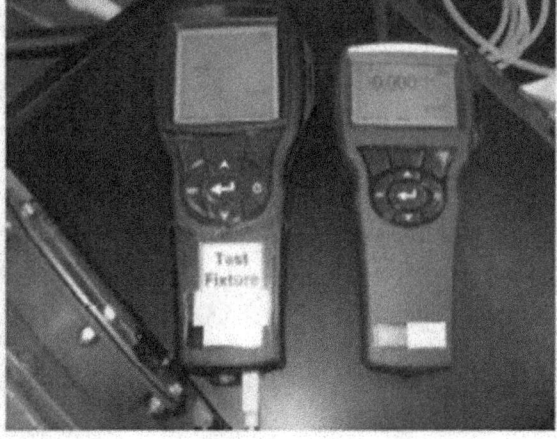

Figure 7.2. Ductwork leakage testing equipment.

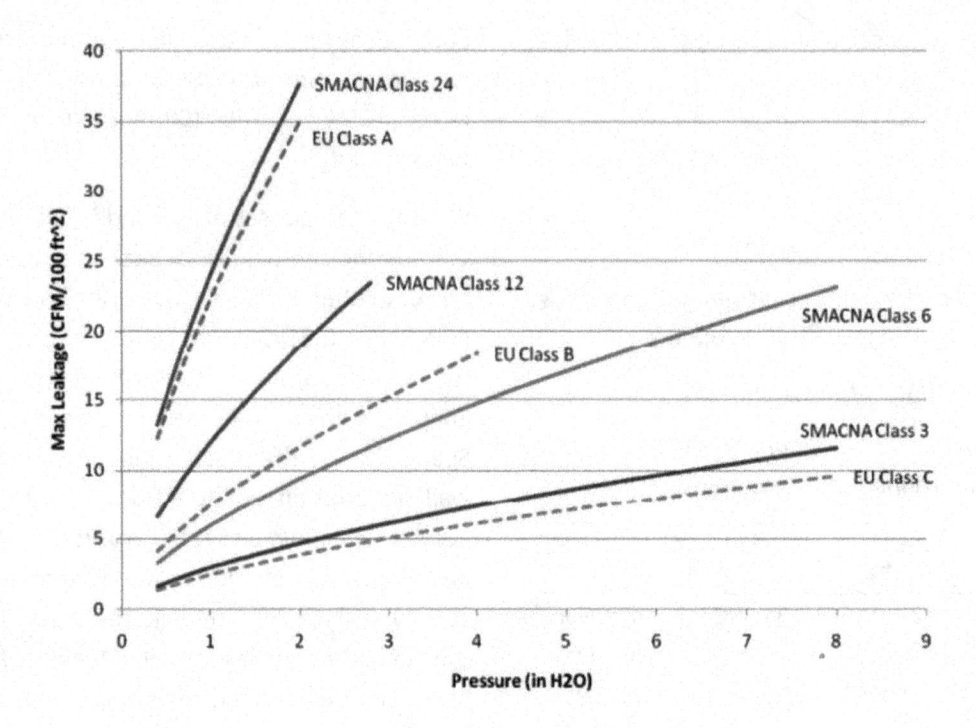

Figure 7.3. Different duct leakage classes as per SMACNA [62] and EN 1507 (rectangular ducts).

- Extraction systems should not be run when building work is ongoing and dirt is present;

- Where a specialist ductwork cleaning company is employed, system commissioning should not commence until cleaning result have been inspected and certified.

- Final checkpoints for ductwork at the end of the commissioning:

- Ductwork is complete;

- Hanging is sufficiently made (min. 2 rods in each point, not too long distances between hanging points) and ductwork has not been bending in straight distances;

- Flexible connectors are installed as specified;

- All joints are sealed properly using aluminium tape or if joints are not properly made, the special tape or spray sealant is used (Fig. 7.4.);

- Insulation is made with appropriate material and is airtight;

- As built drawings have been submitted;

- Ductwork leakage test is completed;

- Ductwork is clean;

- Fire dampers, smoke dampers, volume control dampers (VCD), VAV-dampers and access doors are installed as required.

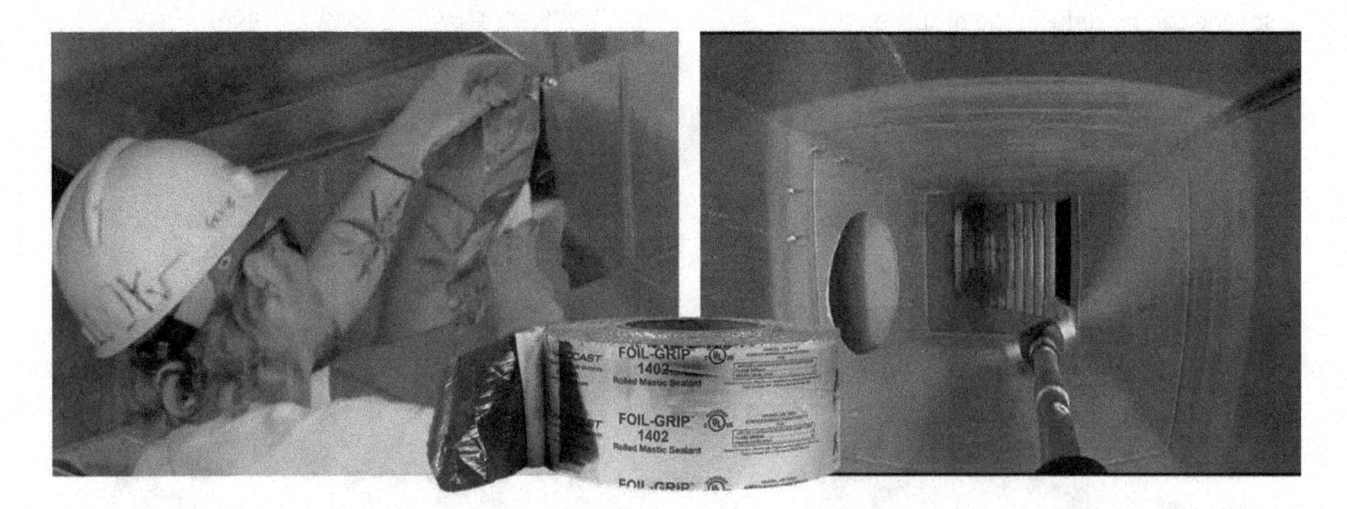

Figure 7.4. There are special duct sealant tapes and sprays available that provides a water-resistant instant bond that seals (up to 10" WG (2,500 Pa) pressure) most surfaces and conforms to surface variations.

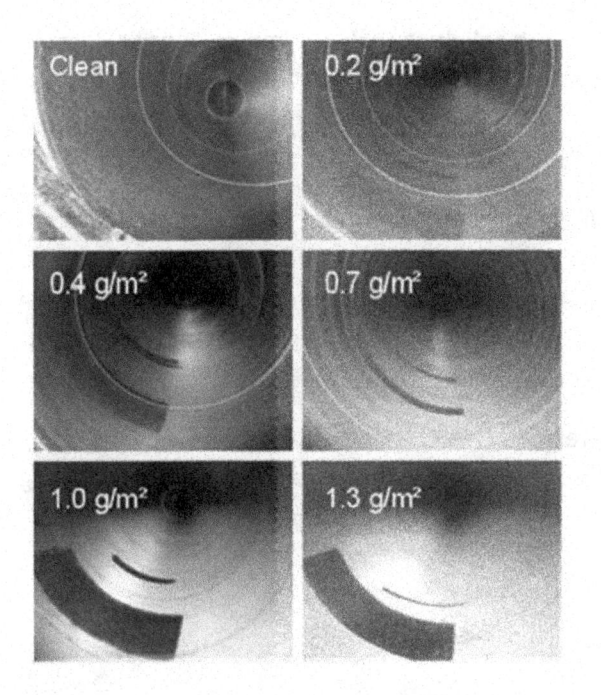 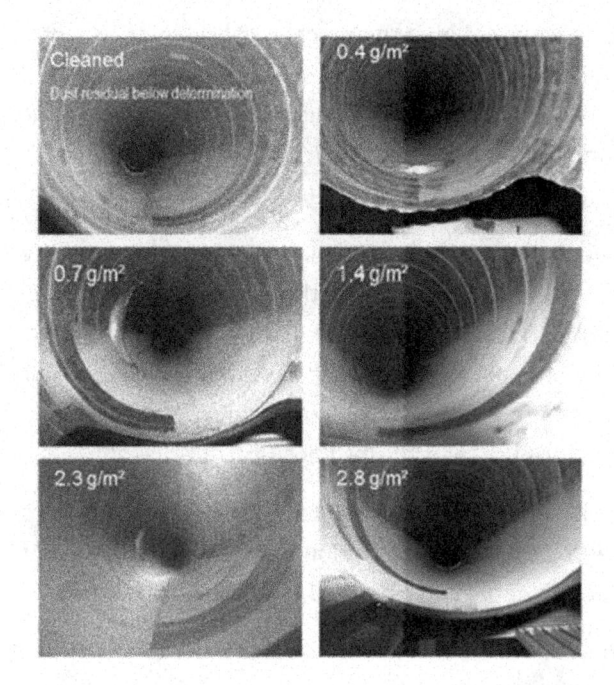

Figure 7.5. Ductwork cleanliness can be evaluated using the visual scale. Left-hand scale is for new ducts and right-hand side scale for existing ducts. [57]

Ductwork cleanliness needs to be ensured during the pre-commissioning audit. Visual inspection of HVAC system components shall be conducted to assess that the HVAC system is visibly clean. An interior surface is considered visibly clean when it is free from non-adhered substances and debris. If a component is visibly clean, then no further cleanliness verification methods are necessary. The pictures presented in Fig.7.5. can be used to estimate the amount of dust in the duct surface.

Another option is to use the NADCA Vacuum Test. It is used for scientifically evaluating particulate levels of non-porous HVAC component surfaces. To be considered clean according to the NADCA Vacuum Test, the net weight of the debris collected on the filter media shall not exceed 0.75 mg/100 cm^2.

In case the duct surface is not clean (dust, grease or microbiological contamination), it needs to be cleaned before air flow balancing.

To ensure that the ductwork cleaning is performed properly e.g. The NADCA Standard for Assessment, Cleaning, and Restoration of HVAC Systems [48] could be followed.

Prior to and throughout the duration of the cleaning process, the HVAC system and associated air duct shall be kept at an appropriate negative pressure differential relative to the indoor non-work area.

This negative pressure differential shall be maintained between the portion of the HVAC duct system being cleaned and surrounding indoor occupant spaces around it.

Service panels are needed to perform assessment, cleaning and restoration procedures. These service panels in the ductwork and respectively trap doors in the ceiling should be included in the designed and built during the ductwork installation. Service panels used for closing service openings in the HVAC system shall be of an equivalent gauge or heavier so as to not compromise the structural integrity of the duct. Panels used for closing service openings shall be mechanically fastened (screwed or riveted) at minimum every 4" on centre. The panel shall overlap the ductwork surfaces by a minimum of 1" on all sides. It is recommended that service panels used for closing service openings be sealed with gaskets, duct sealants, mastic or tape.

HVAC systems shall be cleaned by using a suitable agitation device to dislodge contaminants from the HVAC component surface and then capturing the contaminants with a vacuum collection device. Vacuum collection equipment shall be operated continuously during cleaning. The collection equipment shall be used simultaneously with agitation tools and other equipment to collect debris and prevent cross-contamination of dislodged particulate during the mechanical cleaning process. It is recommended that wet cleaning be performed on air-handling coils, fans, condensate pans, drains and similar non-porous surfaces in conjunction with mechanical methods.

Ductwork should be cleaned before the air flow balancing. However, if the balancing work has already been completed, dampers and any air-directional mechanical devices shall have their position marked prior to cleaning and shall be restored to their marked position after cleaning.

It is recommended that all registers, grilles, diffusers and other air distribution devices be removed if possible, properly cleaned, and shall be restored to their previous position.

Cleaning activities shall not impair, alter or damage any smoke and fire detection equipment located within the facility, or attached to and serving the HVAC system.

When coil cleaning is performed, both upstream and downstream sides of each coil section shall be accessed for cleaning. When both sides of a coil are not accessible for cleaning then removal and/or replacement may be required.

After the cleaning, the video shall be made using the robotic camera in order to ensure the cleaning results in all parts of the ductwork.

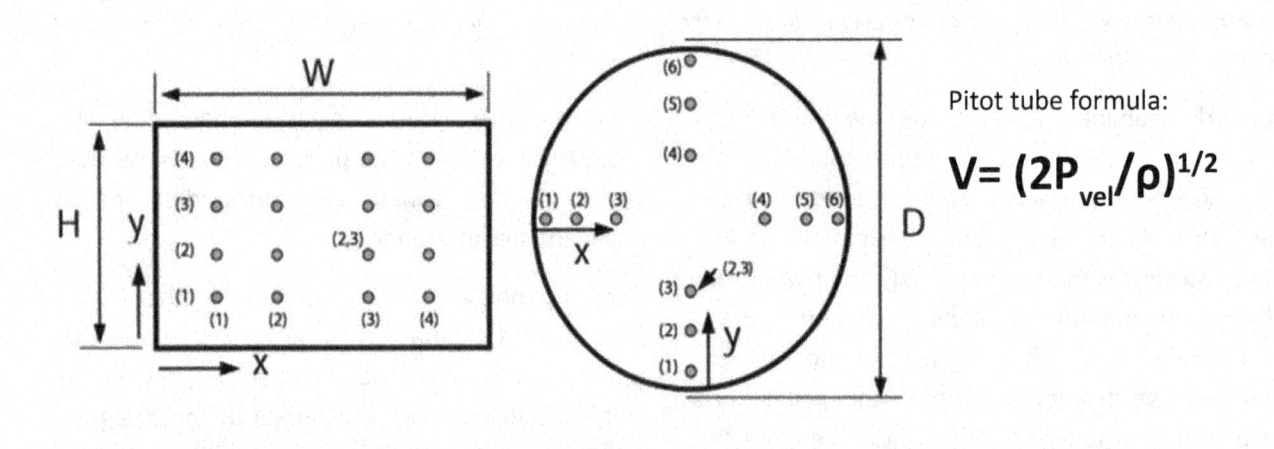

Pitot tube formula:

$$V = (2P_{vel}/\rho)^{1/2}$$

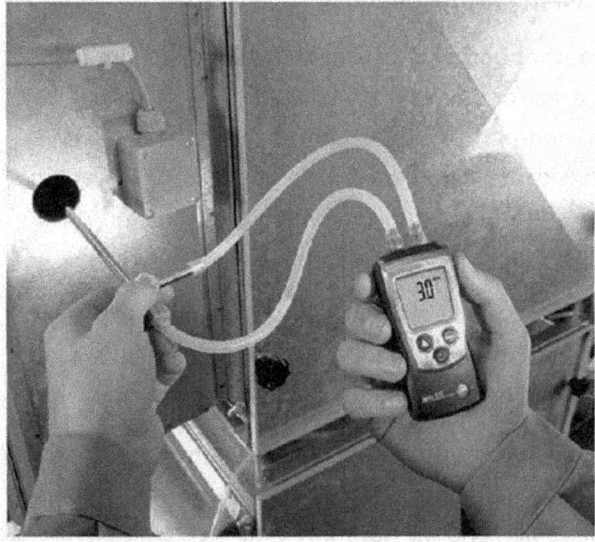

Figure 7.6. As air flow is turbulent inside the duct, there is variation in the velocity over the face area of a duct. This is why the average of several velocity measurement readings gives more accurate result. Air flow measurement can be done either using hot wire anemometer (velocity measurement) or pitot tube (pressure measurement). Typical duct velocities are 1-15 m/s (instrument resolution is typically 0.1 m/s). Pitot tube readings with the same velocities are 2-50 Pa (instrument resolution is typically 1 Pa). Therefore, the accuracy of the pitot tube measurement is poorer with low velocities.

7.2. Ductwork air flow balancing

One of the most important requirements for the design of a duct system is the possibility and simplicity of air flow balancing. The system pressure gets balanced to the point where the fan pressure generation is equal to the sum of the pressure losses through each section of a branch. This is true for each system branch. Another interpretation of the air flow balancing is that pressure losses need to be balanced at each junction. If the sum of the pressure losses in a branch does not equal the fan pressure, the duct system will automatically redistribute air, which will result in air flows that are different from those designed.

Designing a duct system means sizing the ducts and selecting the fittings and fans. Duct sizing is not the same as making pressure loss calculations, although the two are commonly confused. Balancing air systems may be accomplished in various ways.

The most common method to accomplish ductwork balancing is Proportional Method (Fig. 7.7.). [51] Each diffuser is adjusted to supply the right percentage of total air volume in the ductwork. Balancing is performed step by step based on the fact that the operation of each diffuser is dependent on the previous one.

Balancing is made based on the ratio between measured and designed airflow rate. Most of the supply and exhaust ducts can be adjusted based on proportional method.

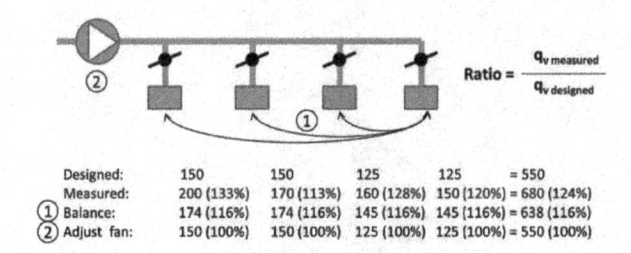

Figure 7.7. Duct balancing principle based on the proportional method.

Proportional method process follows several steps.

Balance the duct work in the following fixed order:

- Ensure that all VCD, VAV and fire dampers are in fully open position;

- Adjust each diffuser damper proportionally to the same ratio in one branch duct;

- Adjust each branch duct VCD damper in every main duct so that each branch duct has the same proportion of air flow rate;

- Adjust each main duct VCD damper in the system;

- After that each diffuser has the same ratio;

- Adjust the fan speed (or airflow by closing the main VCD damper after AHU unit) in the air handling unit:

- Each diffuser has the right airflow rate.

Start the balancing in the main/branch duct that has the highest ratio between measured and designed airflow. If the ratio is higher than 1:3 in the main duct, adjust damper to get ratio to 1:3, but open it before adjusting the main duct.

Duct balancing of each duct branch (Fig. 7.8.) is done based on the following steps:

- Measure the airflow in each diffuser and calculate ratios; (Fig. 7.6 and 7.11.)

- Select reference (last VCD damper of branch) and index (lowest ratio) dampers;

- Adjust reference damper so that the ratio is the same in both reference and index damper (index damper stays fully open);

- Repeat the same with second, third, etc. lowest ratio damper;

- Mark the positions to the dampers and lock them;

- Each branch balancing is completed once all the ratios are the same.

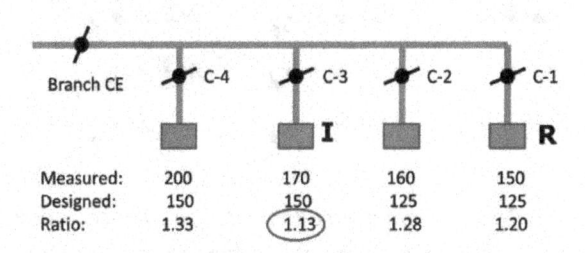

Figure 7.8. Start balancing diffusers by specifying the reference (R) and index (I) dampers. The furthest damper (downstream) becomes a reference damper (called R). Starting point for balancing is the damper that has lowest quotient, i.e. the relationship between measured and designed airflow rate, known as index diffuser (I).

Adjust each branch damper using the similar principle of reference and index dampers. (Fig. 7.9.and 7.10.) Start from the duct with the lowest ratio and adjust the reference damper so that the ratio becomes the same. Hereafter, do not adjust diffuser dampers. Continue balancing in the branch duct damper that has the second lowest ratio.

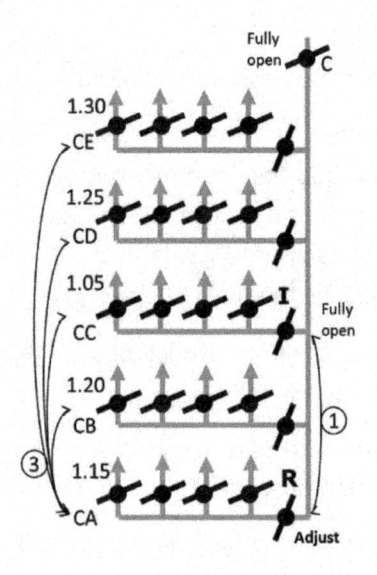

Figure 7.9. Branch duct dampers are balanced based on the same principle starting from the lowest ratio damper and balancing the reference damper against it. And then repeating the same with the damper having the second lowest ratio.

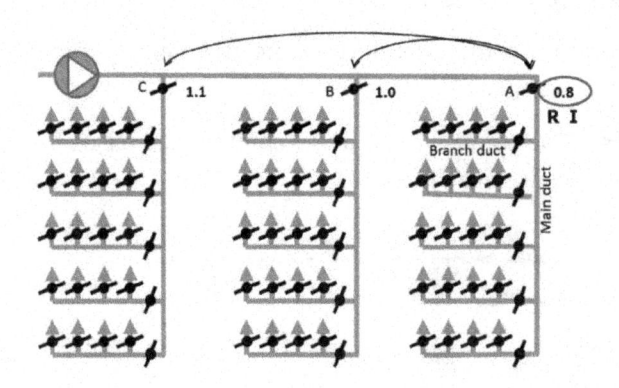

Figure 7.10. The main duct dampers are balanced again based on the same principle. Ratios are specified and reference and index dampers are selected. In this case it is the same damper. This damper A stays in fully open position. Then select the damper with second lowest ratio (damper B) and close that until ratios are the same with Reference damper. Adjust the damper C accordingly.

Finally, the airflow rate in AHU needs to be adjusted:

- Adjust the total air flow rate in air handling unit to the designed value;

- After that the ratio between measured and designed air flow rate should be 1.0 in all diffusers, branch and main ducts;

- Take the following aspects into account:

- Before adjusting the total air flow rate, the building must be in normal operation mode (in design conditions);

- Record the ambient climate conditions;

- Measure the airflow rate after AHU. If that cannot be done in a reliable way, then measure the airflow rate in each main duct;

- Adjust the airflow rate with main damper in AHU or by adjusting the fan speed (recommended);

- Record the amperage and voltage of fan as well as rotating fan speed;

- Measure and record air flow rate of each diffuser and compare it to designed air flow rate. All readings shall be between 90–110 percent;

- Attach the air flow measurement and ratio tables both before balancing and after balancing into the commissioning documents.

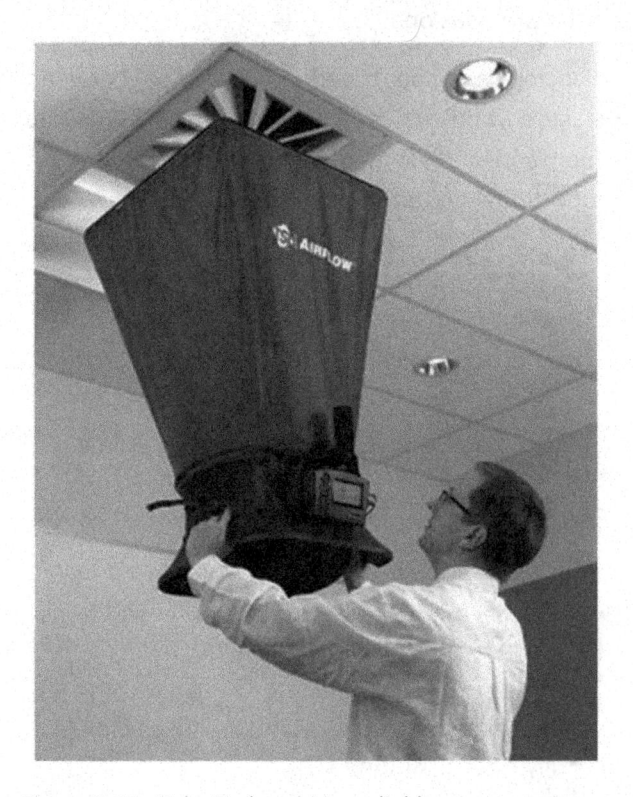

Figure 7.11. Velocity hood is a reliable way to measure the air flow rates in the square diffusers.

7.3. Dedicated outdoor air unit, treated fresh air unit and air handling unit

Air handling unit (AHU) (Fig. 7.12.) is a primary equipment of the air side system which handles and conditions the air, controls it to a required state, and transports it. The basic components of an AHU are:

- Supply and exhaust (DOAS) fan;

- Cooling coil (water or direct expansion (DX) coil);

- Energy recovery wheel (DOAS);

- Air filters;

- Volume control dampers (VCD);

- Controls—like variable frequency driver (VFD), cooling coil controls and measurement sensors;

- Air intake louvre or rain hood with bird screen if installed outdoors.

The air handling unit that supplies the 'fresh' outdoor air for occupants is often called a treated fresh air unit (TFA). It is typically installed into a roof of a building supplying air either directly to the spaces or to the air handling unit rooms. TFA unit typically has air filters, cooling coil for air treatment and supply air fan.

Dedicated Outdoor Air Units (DOAS) that mix outdoor air and return air, are being applied in place of a traditional air handling and treated fresh air unit system. In humid climates, dehumidification requires air to be cooled below the mixed air's dew point even if the occupancy of the building does not require such low temperatures to meet space temperature set points. A better approach for many buildings is consolidating all of the outdoor air treatment into a DOAS unit that supplies 100% outdoor air. Treated air (humidified or dehumidified) from these units can be supplied directly to occupied spaces or can be supplied into the other AHUs dedicated to temperature control. However, some models have an option to return the part of the extracted air to the supply side.

Commissioning is important for all AHUs regardless of size and complexity. AHUs that are not completely commissioned are almost guaranteed to not operate properly. The first issue to be checked is that correct units have arrived on-site. This is recommended when the units arrive on-site but latest during the pre-commissioning.

AHU delivery checklist shall include the following:

- Type of the AHU and fan curve as per design documents;

- AHU tag number on the unit;

- AHU dimensions as per design documents;

- Coil dimensions as per design documents;

- Energy recovery wheel details as per design documents;

- Fan model number as per design documents;

- Motor HP/kWs per design documents;

- Filters as per design documents

- Slope towards drain in the stainless steel drain pan;

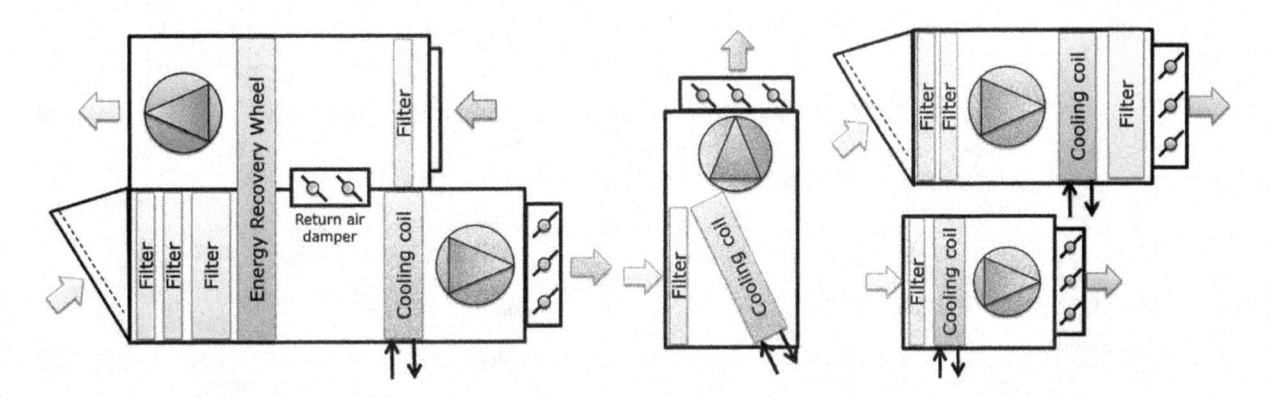

Figure 7.12. Different types of air handling units: draw-through type DOAS unit with return air damper and energy recovery wheel (left), vertical draw-through type air handling unit for indoor installation (middle), blow-through type treated fresh air unit (top right) and draw-through type ceiling suspended unit (bottom right).

Air Filtration and Purification

There is no single technology or filter that can clean all pollutants from the air. Therefore, depending on the application and the current air pollution, the treated fresh air or air handling unit need to have a mixture of different technologies. The following table can be used as a basic guideline when choosing the right set of filters and other cleaning technologies.

Filtration technology	Substances to be removed						Possible side
	RSPM	CO_2	VOC	Gases (NO_2, O_3)	Bio- aerosols	Odour	
Mechanical filters	x						Odour
Electrically charged mechanical filter & ionizer	x				x		Ozone, odour
HEPA filter	x				x		Odour
Electrostatic precipitator	x				x		Ozone
Active carbon filter			x	x		x	Particulates
Potassium permanganate ($KMnO_4$) filter			x	x		x	Particulates
Ultraviolet Germicidal Irradiation (UVGI)					x	x	Ozone
Photocatalytic Oxidation (PCO)			x		x	x	Ozone, gases
Plasma—technology	(x)				x	x	Ozone
Photo-hydro-ionization (PHI)	(x)		x		x	x	Ozone
Ionizer	(x)					x	Ozone, ionized particulates
Ozoniser						x	Ozone
Outdoor air (dilution)	x	x	x	x	x	x	Outdoor pollutants if not filtered

To ensure the proper performance, mechanical filters should be tested as per international standards:

EN 779/Eurovent certified	ANSI/ASHRAE 52.2	ISO 16890			
		ePM_1	$ePM_{2.5}$	ePM_{10}	Coarse
G4/EU4	MERV 8	-	-	-	>80%
M5/EU5	MERV 9-10	-	-	>50%	>90%
M6/EU6	MERV 11-12	-	>65%	-	-
F7/EU7	MERV 13	50-65%	65-80%	>85%	-
F8/EU8	MERV 14	65-80%	>80%	>90%	-
F9/EU9	MERV 15	>80%	>95%	>95%	-

- Canvas connection in the fan outlet;

- Vibration eliminator mounting for fan;

- Orientation of cooling coil connections;

- All factory-build measurement instruments in place;

- All factory-build electrical connections done;

- Volume control damper in SA/RA/FA duct connections;

- Test & Warranty certificates for motor/coil/fan.

The following items need to be checked and confirmed during the AHU pre-commissioning:

- Correct AHU is installed and located as per drawing;

- Clear access to all required components;

- AHU is kept clean inside;

- AHU is free from damage and scratches;

- All transit blocks are removed;

- Supply and return ducts are connected and complete;

- All pipework connections are completed to all coils;

- All panels and filters are installed correctly;

- Temporary filter media is used to cover to protect all filters or temporary filters are used during the commissioning (especially with expensive chemical and fine filters);

- Dampers are free moving through full range;

- Lighting within units is in place and connected;

- Fans and motors rotate properly;

- Fan belts are fitted correctly, pulleys are aligned and belt tension is correct;

- Fan shaft and bearings are aligned;

- Any anti vibration mountings required are installed;

- Motor is properly terminated;

- All electrical isolators are in place and connections complete and tested to motors;

- Power is available for duration of commissioning;

- Static pressure sensor for pressure control is located as specified in the design documents and sensor is correctly installed;

- All heating and cooling coils are secured and fins combed;

- Chilled water pipe connections with supports and valves are done properly;

- Proper condensate drain piping is installed with slopes and supports;

- Chilled water pipe insulation work is done as per design drawing;

- Electrical cabling connections are with cable tray;

- Blanking of gaps are used where necessary with sealant;

- AHU systems must be in operation before the test and balance procedures are started;

- All controls must be installed, calibrated and fully operational;

- Pressure testing of unit has been satisfactorily completed (Tab. 7.2). Factory assembled units are tested on factory and site assembled on site;

- Appropriate safety warning signs are in place;

- All volume control dampers are in fully open position (in case ductwork is not yet balanced).

After the delivery and pre-commissioning checks are complete, the air handling unit start-up can be made. During the start-up several measurements need to be carried out using appropriate methods and instruments (Fig. 6.6., 7.13., 7.14. and 7.15.). All the measurement results shall be recorded and attached to the commissioning documentation.

The operational testing and measurements of air handling units should be carried out in line with the General Measurement and Verification practices and as per the procedures specified below:

- Ensure that the ductwork is properly balanced before making operational measurements in AHU (e.g. as per section 7.2);

- Locate a traverse position in a straight section of each main duct. Refer to the 'Pitot Tube Traverse' procedures;

- Adjust fan speed to obtain 100–110% of design airflow;

- Check the fan for proper operating condition and ensure that the fan motor is below the full-load current;

- Ensure that static pressure control operates properly;

Table 7.2. All AHU units shall be pressure tested to ensure the tightness of unit. In factory-assembled units it is made in the factory before delivery but contractor have to perform the testing with site-assembled units. The tightness criteria can be set e.g. based on the EN 1886 standard.

	A	B	C
Filter Class	G1 – F7	F8, F9	Superior to F9
Negative pressure 400 Pa	L3: 1.32 l/s,m^2	L2: 0.44 l/s,m^2	L1: 0.15 l/s,m^2
Positive pressure 700 Pa	L3: 1.90 l/s,m^2	L2: 0.63 l/s,m^2	L2: 0.22 l/s,m^2
Filter by-pass leakage	6% (G1-G4)/4% (M5-6)/ 2% (F7)	1% (F8)/0.5% (F9)	-
Ductwork Leakage (l/s,m^2)	$0.027 * P^{0.65}$	$0.009 * P^{0.65}$	$0.003 * P^{0.65}$

- Measure the air flow rates in all main ducts and confirm that the total discrepancy is within 10% of design with fan speed of 100 per cent. If not, investigate and balance the ductwork again;

- Determine fan airflow by measuring the air velocity in front of the filter section. Air flow rate should be 100–110% of designed air flow rate. Compare the air flow rate also with the results on the main ducts after the AHU to ensure 0 per cent air leakage in the unit;

- Measure static pressure difference across the fan section (pressure creation) and across all other components in the AHU (pressure loss);

- Check the fan and motor frequency and RPM;

- Measure voltage and current requirements before and after adjustments;

- Measure the operation of energy recovery wheel: air temperatures and flow rates both in supply and exhaust side and calculate the energy recovery efficiency;

- Measure the rotation speed of a wheel and compare it to the selection details;

- After completing the balancing procedure, record the following data:

- Unit design data;

- Nameplate data (must be identified and verified);

- Airflow rates measured from AHU and all main ducts;

- Electrical measurements (voltage and current);

- Motor and fan frequency and RPM;

- Operation point of the fan to the fan curve;

- Drive sizes, belt type, size and number;

- Centre-to-centre distance of the motor-base travel;

- Static pressure profile across each section in AHU according to the static pressure profile procedure;

- Air temperatures across the coils;

- Water flow rates of the water coils and the inlet & outlet temperatures;

- Energy recovery wheel air temperatures across wheel, wheel efficiency and rotating speed.

Measured characteristics must be in accordance with design intent documentation and/or approved submittals.

Air flow rates can be measured using different instruments. Vane anemometer can be used if the air velocity is higher than 0.5 m/s (100 fpm) (Fig. 7.14.) and therefore it is typically used to measure the air flow rate in filter section. Air flow rate in the duct can be measured either using pitot tube or hot wire anemometer (Fig. 7.6.). The measurement principle is the same with all three instruments, i.e. several measurements are taken in the face area and average velocity is calculated. The number of measurement points depends on the face area, but in a 600x600 filter section minimum 5 velocity measurements shall be recorded.

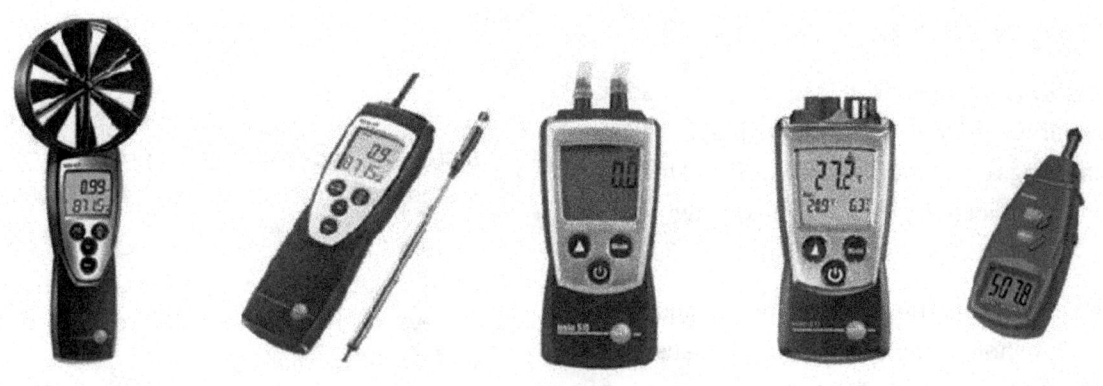

Figure 7.13. Vane anemometer, hot wire anemometer, differential pressure meter, air and surface temperature meter as well as tachometer are required to measure the performance of an air handling unit.

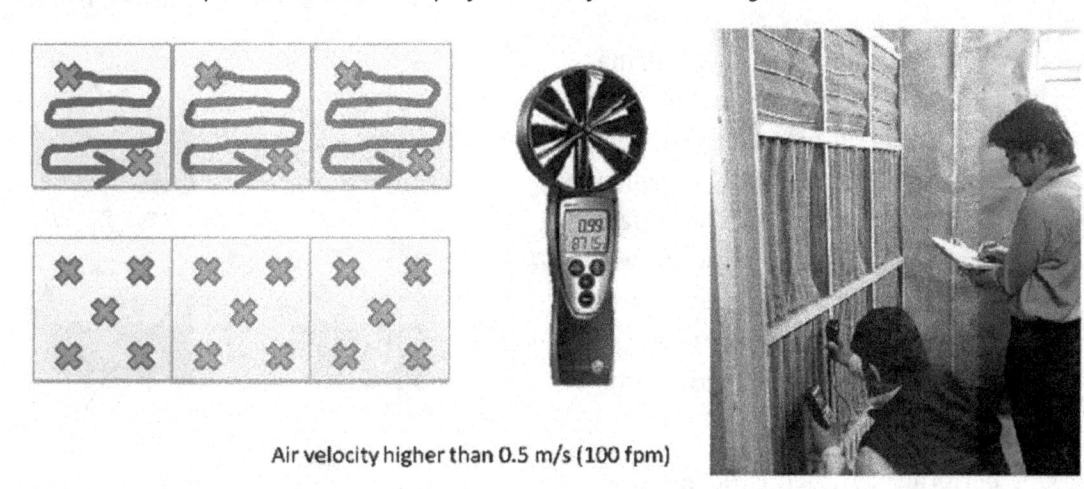

Air velocity higher than 0.5 m/s (100 fpm)

Figure 7.14. AHU air flow rates are typically measured in front of the filter section using the wane anemometer. As the air flow rate is not even in the filter section, several points shall be measured.

Figure 7.15. Operation of each air handling unit needs to be measured during the start-up. All measurement holes need to be sealed after the measurements.

Figure 7.16. Different types of fans are used in ventilation applications. From the left: belt-driven centrifugal fan, centrifugal fan with radial blade impeller, propeller fan and tube-type axial fan.

7.4. Supply and exhaust fans

Single supply and exhaust fan installation and start-up procedures are similar than the fans as a part of an air handling unit. Different type of fans are used for various applications (Fig. 7.16.). The two basic fan types are centrifugal and axial fans.

Centrifugal fans are the most common fans used both in air handling units and as an exhaust fan. Centrifugal fans can then further be classified as radial, forward and backward curved fans. (Fig. 7.17.) Backward curved fans efficiency is better than forward curved blade fans. After reaching the peak power consumption, the power demand drops within backward curved fan's useable airflow range. Whereas, power rises continuously with flow when using forward curved fans. This also leads to the fact that it is more expensive to operate forward curved fans even they their first-cost is lower. Backward-inclined fans are non-overloading' because the changes in static pressure does not overload the motor.

Axial-flow fans performance again is different, as its produce less pressure than centrifugal fans, and the pressure they generate dips before reaching the peak pressure point. They are typically equipped with adjustable/variable pitch blades to meet varying flow rates at low pressure, whereas tube-axial fans have medium pressure and high flow capability.

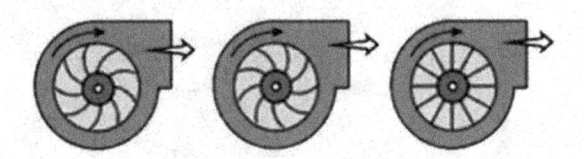

Figure 7.17. Forward curved, backward curved and radial fan blades.

The fan operation point is defined in a fan curve. During the design review it is important to ensure that the fan curve of a selected fan is sufficient to each application. At the end of the start-up process, the actual fan operation point needs to be recorded and compared with the design point (air flow rate and pressure). Note that in case the fan speed is higher than in design conditions, the energy use of the fan is also higher than expected. Also a higher fan speed may lead to noise problems.

There are three options to adjust the air flow rate:

- Increase or decrease the system curve, i.e. increasing or decreasing the pressure loss of the system by adjusting the volume control damper.

- In V-belt driven systems when the air flow rate change is permanent, a drive or driven or both pulleys can be changed to increase or decrease air flow rate.

- Increase or decrease of fan speed allows the continuous and energy efficient control of air flow rates. (Fig. 7.18. and 7.19.)

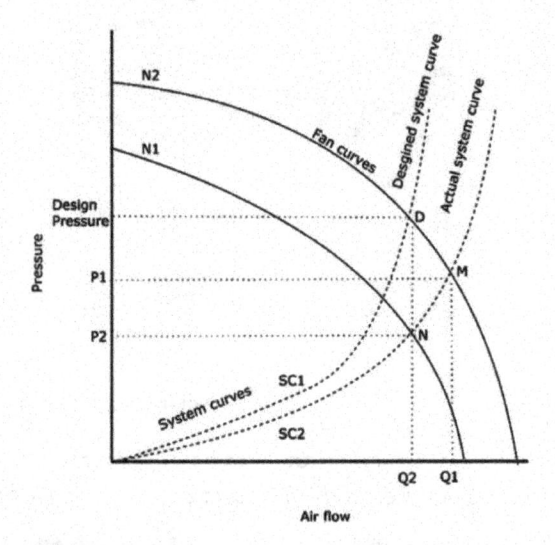

Figure 7.18. In this case (point M) the pressure loss of system is higher than designed and therefore the designed air flow rate (point D) cannot be achieved. The solution is to increase the fan speed to achieve the required air flow rate (point N).

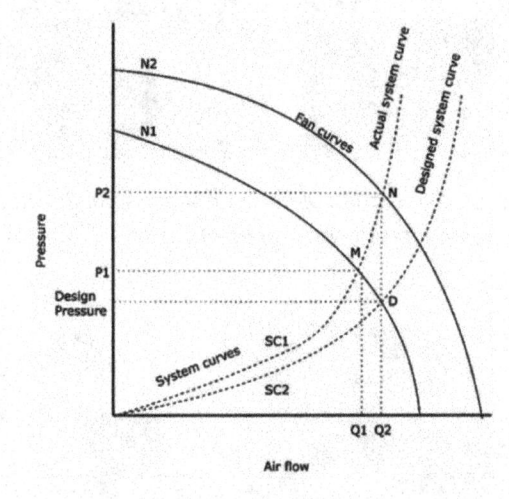

Figure 7.19. There are two ways to reduce the air flow rate from Q1 (point M) to Q2. The first option is to change the system curve from SC1 to SC2 by closing the VCD damper after the fan (point D). The second option is to reduce the fan speed (point N).

The fans operate under a predictable set of laws concerning speed, power and pressure. A change in fan speed (RPM) of any fan will predictably change the static pressure and power necessary to operate it at the new fan speed:

$Q_1/Q_2 = N_1/N_2$, (Q = air flow rate, N = fan speed)

i.e. 10% increase in fan speed increases 10% of air flow rate.

$SP_1/SP_2 = (N_1/N_2)^2$, (SP = static pressure)

i.e. reduction of fan speed by 10% decreases the static pressure by 19% and an increase in fan speed by 10% increases the static pressure by 21%.

$P_1/P_2 = (N_1/N_2)^3$, (P = power in kW)

i.e. reduction of fan speed by 10% decreases the power requirement by 27% and an increase in fan speed by 10% increases the power requirement by 33%.

It is important that the correct fans arrive at the construction site. Therefore, the following items need to be checked during the delivery of fans:

- Condition of equipment;

- All required accessories are installed;

- Equipment data is in compliance with specifications and drawings;

- Performance curve of fan is appropriate for use.

Pre-commissioning ensures that the fan is ready for start-up. The following checks shall be carried out:

- Inspect all flexible fabric connections between fans and ductwork to ensure that a fabric 'bellows exists when fans are operating;

- Check the internal and external cleanliness of fans;

- Check that all components, bolts, fixings, etc., are secured;

- Make sure that the impeller is secured and it is free to rotate;

- Check fan impeller and motor shaft alignment (Fig. 7.20.);

- Fans are installed for correct airflow direction and, when compounded, in correct order.

- Check the level and plumb of fan, motor shaft and slide rails;

- Ensure anti-vibration mountings for correct deflection and the removal of transit bolts and packing materials;

- Check the static balance;

- Ensure that correct VFD drive is fitted;

- Check securing and alignment of pulleys and couplings;

- Check belt tension and match (Fig.7.21.);

- Check cleanliness of the bearings, that the lubricant is fresh and of the correct grade;

- Ensure that coolant is available at the bearings when specified;

- Check that drive guards are fitted and access for speed measurement is provided;

- Ensure satisfactory operation of the inlet guide vanes over full range of movement;

- Check that fan casings are earthed correctly and soundly bonded.

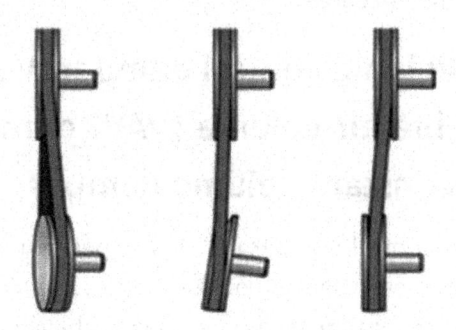

Figure 7.20. In belt-driven fans it is important that the impeller and motor shafts are parallel in both the vertical and horizontal planes and that sheave are both in axial and radial alignment.

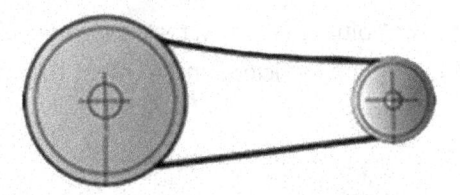

Desired belt deflection = belt span/64.

Figure 7.21 Belts must have the proper tension. Belts that are either too loose or too tight cause vibration or excessive wear.

Operation checks and measurement includes:

- Measure and adjust the air flow rate by adjusting the fan speed to be between 90–100% from designed value;

- Measure the pressure difference across the fan (head);

- Check the fan and motor frequency and RPM;

- Measure voltage and current requirements before and after adjustments;

- Measure and record equipment vibration, bearing vibration, equipment base vibration and on building structure adjacent to equipment:

- For fan bearing: drive and opposite end;

- Motor bearing: drive and opposite end;

- Equipment base: top and side within 6" of each isolator;

- Building: Floor adjacent to fan/motor, within 6" of each isolator:

- Record the air flow rate, static pressure difference and mark the operation point of fan to the fan curve.

7.5. Volume control damper (VCD), Variable air volume (VAV) damper and Constant volume damper

Volume Control Damper (VCD) is a specific type of blade damper used to control the flow of air in a ventilation system. They are used to balance the correct air flow rates to different spaces. VCD can be rectangular, multi-blade damper or circular single blade or iris-type damper. It is typically operated manually. (Fig. 7.22.)

Variable Air Volume (VAV) systems were developed to be more energy efficient and to meet the varying heating and cooling needs of different building zones. Room thermostats control the amount of primary air delivered to each zone through VAV dampers. These dampers vary the volume of air to each zone according to the cooling needs. (Fig. 7.23.)

As the air volume for the zones varies, the static pressure in the main duct varies too. A static pressure sensor in the main duct controls the fan speed to maintain a constant supply duct static pressure.

VAV systems are either pressure dependent or pressure independent. In pressure dependent systems, the volume of air supplied by the terminal unit varies depending upon the static pressure in the primary air duct. Pressure independent terminal units have flow-sensing devices that limit the flow rate through the box. They can control the maximum and minimum air flow rate that can be supplied and are therefore independent of the static pressure in the primary air duct.

Constant volume dampers (CVD) are used in those branches of VAV system, where air flow rate shall be constant regardless duct static pressure (Fig. 7.24.)

The following issues need to be checked during the pre-commissioning of dampers:

- Make sure that the location in the ductwork allows proper air flow control operation:

- Select the installation location for the VAV controller such that the control components remain accessible and there is enough clear space at the inspection accesses;

- Be sure to install the unit according to the airflow direction indicated by the arrow on the unit;

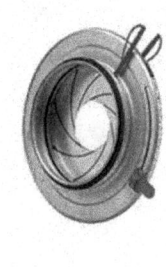

Figure 7.22. Square multi-blade volume control damper, circular iris-type volume control damper and VAV-damper.

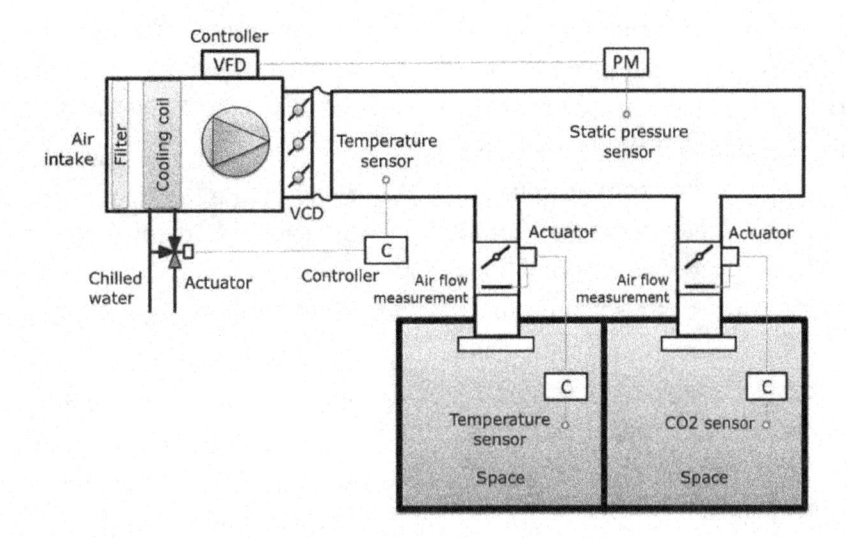

Figure 7.23. Operation schematic of VAV damper. Room air temperature or CO_2 is controlled by adjusting air flow rate with VAV-damper. Static pressure sensor of the fan shall be located 2/3 from the duct length, not directly after the fan, to ensure proper pressure management.

When installing elbows, flaps or other disruptive bodies, provide a straight duct section of 3–5D upstream;

- Ensure that connections are airtight;

- Secure VAV controllers with screw or rivet connection onto both spigot connections to prevent rotation;

- Connect the onward ducting to the VAV controller in an electrically conductive connection by means of ground straps on both ends in order to provide the necessary equipotential bonding;

- Connect the duct to the building's equipotential bonding;

- Ensure that the connections for the supply voltage and signal lines of the actuator are made in the terminal box;

- Observe damper motors and actuators through an operating cycle to check for defects or binding;

- Linkages from actuators should be adjusted to ensure that the blades of damper fully opens or closes within the stroke or travel of the actuator arm;

- Blades should be checked in closed position to be sure they all close tightly. If necessary, adjustment should be made to damper linkages to close any partially open blades;

- Damaged blades should be replaced. Dirt, soot, lint, etc., should be removed, especially around operating parts;

- Check blade edge and side seals and replace wherever necessary;

- Check pins, straps, bushings (bearings) for wear, rust or corrosion and replace, as required;

- Check the lubrication of all mechanisms and moving parts;

- Caulking, used to fix frames to the structure, should be checked and repaired, as needed.

After the start-up of all ventilation system components, the operation check of VAV damper shall be carried out, including the following:

- Set the minimum (e.g. 30% of maximum air flow rate) and maximum air flow rates as per manufacturer's instructions;

- Measure air flow rate and duct velocity (to ensure that damper is in operation area) in fully open position as well as in minimum operation point;

- Record the air flow rates and positions of damper into commissioning documentation.

Figure 7.24. There are two basic models of system powered constant volume dampers: one to be installed inside the duct (left) that has typically only one preselected air flow rate and the other one (right) in which the air flow rate can be re-adjusted.

In case system powered constant volume dampers are used to control the air flow rates, proportional balancing of ductwork is not required, but damper is set to the correct air flow rate as per manufacturer's instructions. After commissioning, the static pressure across the damper is measured in order to ensure that product is on its operation area (typically 50–1000 Pa). It is also important to ensure that damper location is sufficient and enough straight duct is before damper (>1.5 D). After commissioning, record the damper positions and measured static pressures and attach them to the commissioning documents.

7.6. Room air diffusion

Room air diffusion impacts both thermal comfort of occupants and indoor air quality. In case air jet from the diffuser reaches occupied zone with too high velocity, users may feel a draught. This is often the case if the supply air is very cold or jet is too long (too small or the wrong type of diffuser) when colliding with another jet or an obstacle. Also the wrong positioning of the grille may create a local draft if the jet is not detached properly to the ceiling, or the grille blows straight to the occupied zone. The sufficient diffuser size and type for each condition can be selected based on manufacturer's data of jet throw (Fig.7.25):

- Throw length L0.2 (distance where jet velocity is 0.2 m/s (40 fpm)) shall be 0.9 - 1.4 times the distance between the diffuser and the wall, depending on the application;

- The minimum distance L_{min} between diffusers is defined based on the air velocity of combined jets reaching the occupied zone;

- Diffuser size, air flow rate and supply air temperature defines the distance of jet detachment Ld (point where jet detach from ceiling surface). The colder the supply air, the shorter the L_d.

Also, the location of return air too is important. If it is too close to the supply air diffuser, the fresh air may not circulate in the space but is exhausted directly (short circuit). This also reduces the ventilation efficiency in the space.

If the air jet is too short to reach the occupied zone or there are heavy thermal plumes that turn the jet back to the upper part of the room, then air quality may be compromised. Therefore, it is important that supply air reaches the entire occupied zone without creating too high velocity.

A Displacement Ventilation (DV) system comprises cool air supplied at low velocity from a low level wall -mounted or floor-mounted ventilation terminal, which is located within the occupied space. The supplied cool air initially remains at floor level, while it moves across the space. Natural convection resulting from internal heat sources such as occupants and equipment induces upward movement of this air within the occupied space. The warm air that collects over the occupied space is then extracted by a mechanical extract system or by natural means.

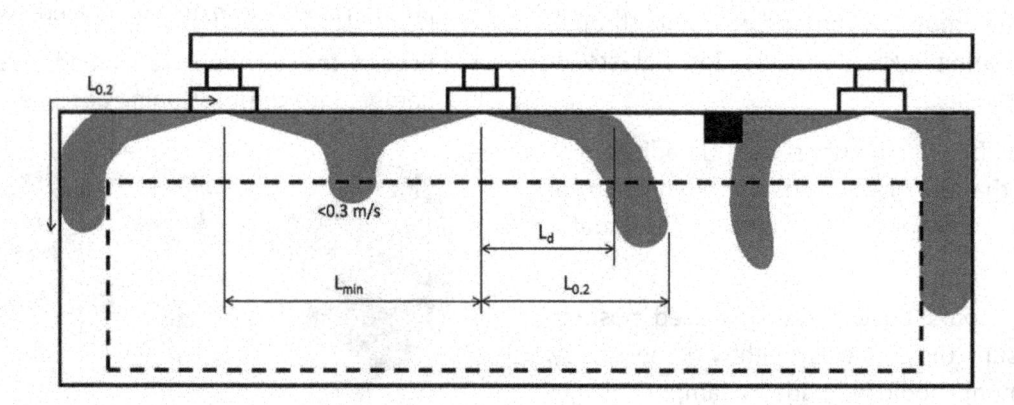

Figure 7.25. Room air diffusion impacts the occupied zone velocity conditions as well as the indoor air quality. Therefore, air shall be diffused to the space as evenly as possible. [48]

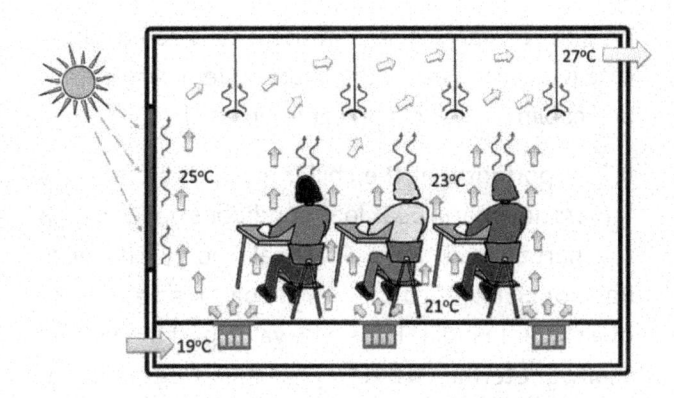

Figure 7.26 Principle of underfloor air diffusion (UFAD) system.

Under Floor Air Diffusion (UFAD) is a method of delivering cooled/conditioned air using the open space (under-floor plenum) between the structural concrete slab and a raised access floor (Fig. 7.26). Air is directly delivered into the occupied space of the building. Air can be delivered through different types of supply outlets located at floor level. Supply outlets are typically swirl type outlets which mix the cold supply air effectively with the indoor air near the diffuser. In auditoriums, where diffusers are placed under the seats, low velocity diffusers are used. There are three basic approaches to configuring the supply-air side of an UFAD system:

- Pressurized underfloor plenum with a central air handling unit delivering air through the plenum and into the space through passive grills/diffusers;

- Zero-pressure plenum with air delivered to the space through local fan-driven supply outlets in combination with the central air handling unit. However, operation of fans reduces the energy efficiency of this approach;

- Supply air can also be ducted through the under floor plenum to each supply outlet.

In Displacement Ventilation or in Underfloor Air Diffusion (UFAD), there is a risk of too high velocity near the floor, especially if the supply air is too cold. In these applications, the air velocity measurements are recommended near the floor. If the diffuser is near the heat source, e.g. a warm window, it is important to ensure that not all of supply air rises near the heat source, but it ventilates evenly in the entire space.

During the pre-commissioning, make sure that:

- The diffusers and grilles are properly installed;

- The diffuser size and distance between the diffusers are sufficient in order to avoid too high velocities in the occupied zone (too close to each other) but also ensure that the entire occupied area is properly ventilated;

- Ensure that there are no obstacles near the diffuser to disturb the throw pattern;

- In case of DV, make sure that the air is supplied evenly in the entire face area oflow velocity displacement terminal unit;

- In case of UFAD, ensure that diffusers are correctly installed and the plenum is tight enough for this application, i.e. air comes to the space via diffusers and outdoor air is not leaking into a plenum.

After the start-up of all ventilation system components, the diffuser operation needs to be studied as following:

Measure and record the air flow rate in each diffuser using velocity hood or measuring the air flow rate from the duct (using hot wire anemometer or pitot tube) - vane anemometer readings from diffuser or grille surface are not accurate;

Study the throw pattern to ensure correct velocity conditions in the occupied zone by velocity measurements or visualizing jet using smoke;

In case of DV and UFAD, measure the air temperatures and velocities near the floor and make sure that comfort criteria are met;

In case of VAV system, measure the air flow rate and study the throw pattern of diffusers with different VAV-damper positions;

In demanding applications, the tracer gas measurement can be used to study the ventilation efficiency in different parts of the space;

Record and attach all measurement results to the commissioning documents.

7.7. Chiller

A chiller is a machine that removes heat from a liquid, typically via a vapor-compression (Fig. 7.27 and 7.30.), absorption or adsorption refrigeration cycle. This liquid can then be circulated through a heat exchanger to cool air or equipment as required. As a necessary by-product, refrigeration creates waste heat that must be exhausted to ambient or, for greater efficiency, recovered for heating purposes. [73]

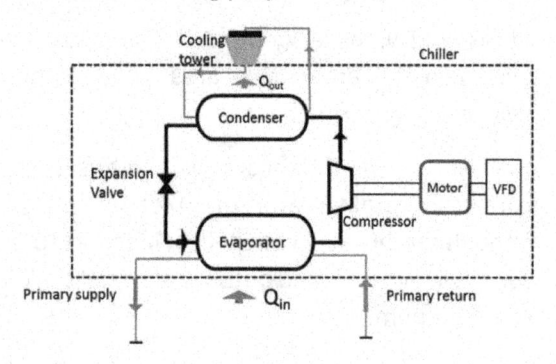

Figure 7.27. A vapour compression chiller consists of four major components: a compressor, an evaporator, a condenser and an expansion valve.

Each chiller has unique characteristics that affect full -load and part-load efficiency. These include compressor design, evaporator and condenser heat transfer characteristics, unloading devices (such as variable speed drives, slide valves and inlet guide vanes), oil management systems, and internal control logic.

Chiller efficiency can be specified using different performance indicators.

The Coefficient of Performance (COP) of a chiller is specified with the following formula:

COP = Cooling Power \ Input Power

Vapour compression chiller's coefficients-of-performance (COPs) is very high, typically 4.0 or more depending on the type of the chiller, refrigerant and application. COP of absorption and adsorption chillers are much lower, often just above 1.0, but the efficiency of cooling system typically comes from the fact that they are operated with free energy source (e.g. waste heat or solar heat).

Another commonly used indicator of efficiency is 'kW/TR'. It describes how many kW of electricity is required to produce 1 refrigerant TON of cooling. COP of 8 is equal to 0.44 kW/TR and COP of for is equal to the 0.88 kW/TR. In case the efficiency is for the chiller

only, the compressor energy use is considered, but in case it is calculated for the entire system, also pumps and cooling tower fan power is included.

It is important that the chiller meets the efficiency targets not only in peak load conditions but also with the partial load. India Seasonal Energy Efficiency Ratio or ISEER [41] describes the part load efficiency of a chiller in India. The single value part load rating shall be determined by using the following equation:

ISEER = 6 * COP 100% + 48 * COP 75% + 36 * COP 50% + 10 * COP 25%

where:

- COP100% = COP at full load rating point and operating condition.
- COP75% = COP at 75% load rating point and operating conditions.
- COP50% = COP at 50% load rating point and operating conditions.
- COP25% = COP at 25% load rating point and operating conditions.

The 'weighting factors' for each load rating point are based on the weighted average of the most common building types across climatic zones of India. During the commissioning, it is important that a chiller is chosen based on ISEER, not any international efficiency rating.

Water-cooled centrifugal chillers for comfort-cooling applications are generally designed for a set of standard conditions e.g. a leaving chilled-water temperature (LCHWT) of 44 °F (6 °C) and an entering condenser-water temperature (ECWT) of 85 °F (29 °C). Chiller lift (Fig. 7.28.) is defined as the difference between the refrigerant saturated condensing and evaporating pressures. Using defined pressure-temperature relationships, lift can also be measured with the leaving chilled water and condenser water temperatures. Further, when the LCHWT and

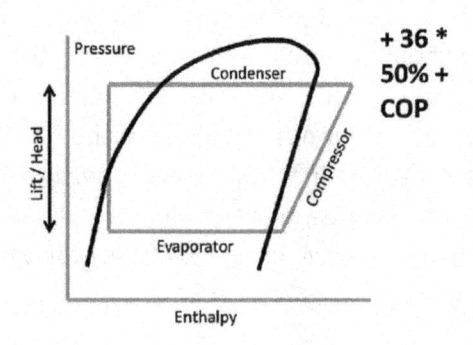

Figure 7.28. Chiller lift is the evaporating and compressing pressure difference.

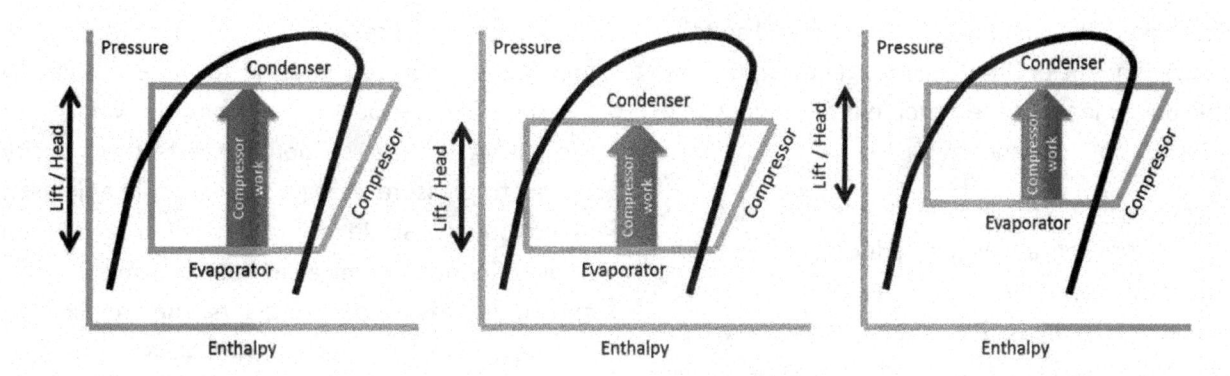

Figure 7.29. Chiller lift can be reduced and energy efficiency of cooling system increased by reducing the lift. This can be done by reducing the entering condenser water temperature (ECWT) or by increasing the leaving chilled water temperature (LCHWT).

condenser-water flow is constant, the ECWT can be used as a metric for lift.

Lower chiller lift improves chiller efficiency. (Fig. 7.29.) Lift can be reduced either by lowering ECWT or by increasing the LCHWT. In comfort-cooling applications, ambient weather conditions may allow facility owners to take advantage oflower ECWT (in cold and temperate weather conditions). In high outdoor air conditions, increasing the LCHWT is easier to implement. The sufficient dehumidification of humid outdoor/indoor air in the ventilation and AC systems' design need to be taken into account when choosing the LCHWT. However, not every are commonly seen in automotive applications. chiller is designed to take advantage of conditions Larger reciprocating compressors, well over 750 kW when high LCHWT is specified. There are four design variables that affect a centrifugal chiller's ability to handle low-lift conditions encountered in process cooling applications:

- The drive design;
- The orifice design;
- The oil management system;
- The compressor aerodynamics.

In a fixed speed centrifugal chiller, inlet guide vanes are used to throttle refrigerant at the inlet of the compressor to vary the compressor capacity. When the guide vanes move from fully open to close, the vane action eventually becomes a restriction, losing the pre-rotation effect and reducing the compressor capacity and efficiency.

Variable frequency drive slows the compressor to match the head conditions, without closing the guide vanes. Variable frequency drive has to continuously monitor several operating conditions such as chilled water temperature, chilled water set point, refrigerant pressures, pre-rotation vane position and the actual

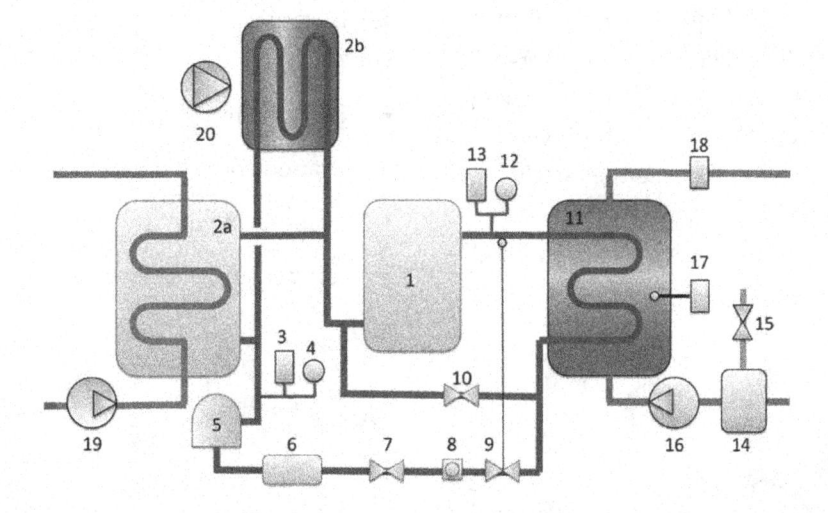

Figure 7.30. Chiller components: 1) compressor, 2a) water cooled condenser, 2b) air cooled condenser, 3) high pressure limit, 4) high pressure gauge, 5) liquid receiver, 6) filter drier, 7) liquid line solenoid, 8) refrigerant sight glass, 9) expansion valve, 10) hot gas bypass valve, 11) evaporator, 12) low pressure refrigerant gauge, 13) low pressure refrigerant limit, 14) water reservoir 15) water makeup solenoid, 16) water pump, 17) freezing limit, 18) evaporator flow switch, 19) cooling tower pump and 20) air cooled condenser fan.

motor speed. It optimizes both motor speed and the pre-rotation vane position and changes the frequency of the power input to the motor in order to consume the least amount of energy.

A compressor is a mechanical device that increases the pressure of a gas by reducing its volume. Compressor operation as it exists in most AC applications is based on vapor compression. The compression principle (e.g. reciprocating/scroll/ screw/centrifugal) impacts the efficiency of a chiller and the application of where to use it. (Fig. 7.32).

Reciprocating compressors use pistons driven by a crankshaft. They can either be single or multi-staged, and can be driven by electric motor or internal combustion engines. Small reciprocating compressor are commonly seen in automotive applications. Larger reciprocating compressors, well over 750 kW are commonly found in large industrial applications. Discharge pressures can range from low pressure to very high pressure (200 MPa). [73]

A scroll compressor uses two interleaved spiral-like vanes to pump or compress fluids such as liquids and gases. The vane geometry may be involute, Archimedean spiral or hybrid curves. They operate more smoothly, quietly, and reliably than other types of compressors in the lower volume range. Often, one of the scrolls is fixed, while the other orbits eccentrically without rotating, thereby trapping and pumping or compressing pockets of fluid between the scrolls. [73].

Screw compressors (Fig. 7.38) use two meshed rotating positive-displacement helical screws to force the gas in to a smaller space. These are usually used for continuous operation in commercial and industrial applications and may be either stationary or portable. Their application can be from 2 kW to over 900 kW and from low pressure to moderately high pressure (10 MPa). [73]

Centrifugal compressors use a rotating disc or impeller in a shaped housing to force the gas to the rim of the impeller, increasing the velocity of the gas. A diffuser section converts the velocity energy to pressure energy. They are primarily used for continuous, stationary service in industries such as oil refineries, chemical and petrochemical plants and natural gas processing plants. Their application can be from 75 kW to >1000 kW. With multiple staging (Fig. 7.31.), they can achieve high output pressures greater than 70 MPa. [73]

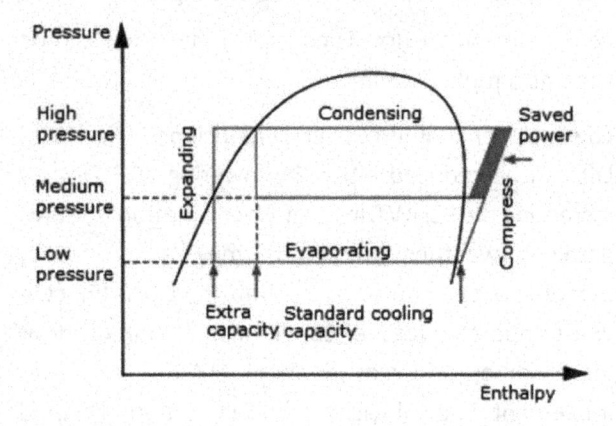

Figure 7.31. 2-stage compressing can reduce the compression ratio to lower down the evaporating temperature, save compressing power and increase the system COP.

An expansion valve (Fig. 7.33. and 7.37.) is a component in refrigeration systems that controls the amount of refrigerant flow into the evaporator, thereby controlling the superheating at the outlet of the evaporator. The thermostatic expansion valve can be selected when the following are known:

- Refrigerant;
- Evaporator capacity;
- Evaporating pressure;
- Condensing pressure;

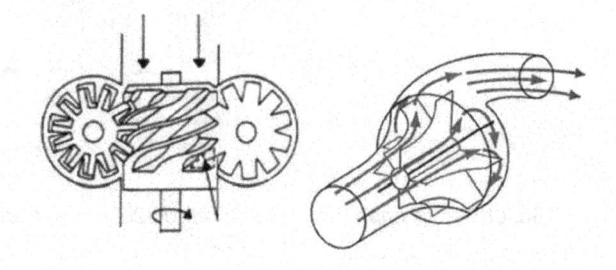

Figure 7.32. Compressors in vapour compression chillers can be based on the different principles of compression - from the left: reciprocating, scroll, screw-driven and centrifugal compression

- Sub-cooling;
- Pressure drop across valve;
- Internal or external pressure equalization.

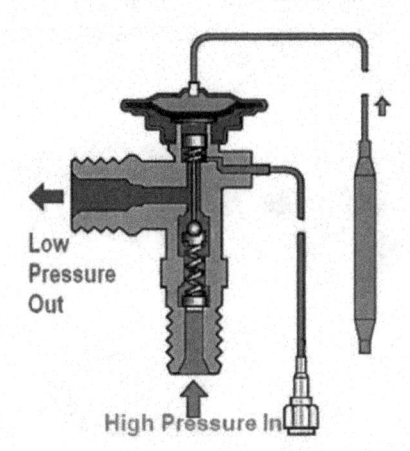

Figure 7.33. Thermal expansion valve consists of valve body, sensing bulb and in some models an external pressure equalizer.

Superheat (Fig. 7.34.) is measured at the point where the sensing bulb is located on the suction line and is the difference between the temperature at the bulb and the evaporating pressure/evaporating temperature at the same point. Superheat is used as a signal to regulate liquid injection through the expansion valve.

Sub-cooling is defined as the difference between condensing pressure/temperature and liquid temperature at the expansion valve inlet. Sub-cooling of the refrigerant is necessary to avoid vapour bubbles in the refrigerant ahead of the expansion valve. Vapour bubbles in the refrigerant reduce capacity in the expansion valve and thereby reduce liquid supply to the evaporator. Sub-cooling of 4–5 K is adequate in most cases.

Expansion valves with external pressure equalization must always be used if liquid distributors are installed or with small compact evaporators, e.g. plate heat exchangers.

The expansion valve must be installed in the liquid line, ahead of the evaporator, with its bulb fastened to the suction line as close to the evaporator as possible. If there is external pressure equalization, the equalizing line must be connected to the suction line immediately after the bulb. The bulb is best mounted on a horizontal suction line tube. Its location depends on the outside diameter of the tube.

An evaporator is used in an air-conditioning system to allow a compressed refrigerant to evaporate from liquid to gas while absorbing heat in the process. It is a heat exchanger that transfers heat from the substance being cooled to a boiling temperature.

When subcooled liquid refrigerant at high pressure state (a) expands through the expansion valve, the pressure and thus the saturation temperature both decrease (b). The amount of flash gas formed after the expansion valve decreases with the level of sub-cooling and the evaporator inlet pressure. The mixture of liquid and gas from the expansion valve enters the evaporator and starts to boil, because heat is transferred from the warmer secondary fluid (b-c). The evaporating refrigerant absorbs energy from the secondary fluid, whose temperature is reduced. After full evaporation, when 100% of the refrigerant has become saturated vapor (c), the temperature of the vapor will start to increase, i.e. the vapor will become superheated. The refrigerant flow leaving the evaporator will be 100% superheated vapor (d).

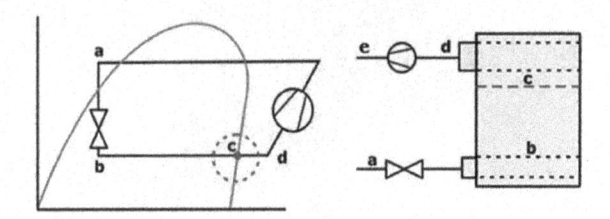

Figure 7.34. The mixture of liquid and gas from the expansion valve enters the evaporator and leaves as superheated vapour.

Two major categories of evaporators (Fig. 7.35.) are used in buildings: air cooled coils and liquid cooling coils. Air cooled coils are used in room AC units and as a direct expansion (DX) cooling coil in the AHU. Good thermal contact between the fins and tubes is a must to ensure efficientheat transfer. The spacing of the fin depends on the operating temperature of the coil. In air conditioning application, 14–16 fins per inch may be used as long it is designed in such a way that frost does not accumulates in the coils. Too many fins reduce the capacity of the evaporator by restricting the flow over the heat exchanger.

Liquid cooling coils are used to cool the chilled water that is supplied to AHU water cooling coil or to room cooling units. Liquid cooling evaporators can be direct expansion or flooded type.

A flooded evaporator operates in conjunction with a low-pressure receiver. The receiver acts as a separator of gaseous and liquid refrigerant after the expansion valve and ensures a feed of 100% liquid refrigerant to the evaporator. Unlike in a direct expansion (DX) evaporator, the refrigerant is not fully evaporated and superheated at the flooded evaporator outlet. The leaving refrigerant flow is a two-phase mixture with typically 50–80% gas.

There are three main factors to consider when designing an evaporator:

- Pressure Drop between the outlet and the inlet needs to be low enough to ensure high enough refrigerant circulation;

- Temperature: the evaporator must have enough surface to absorb the required heat;

- The evaporator must have enough space for the liquid refrigerant and the vapor to separate from the liquid.

Figure 7. 35.. Two different kinds of evaporators that are used in HVAC applications: direct expansion (DX) coil and flooded liquid cooling coil.

A condenser's (Fig. 7.36.) function is to transform hot discharge gas from the compressor to a slightly sub-cooled liquid flow, by transferring heat from the refrigerant gas to the secondary cooling liquid. The basic operation of condensers is divided into three parts: de-superheating, condensation and sub-cooling. All three operations can be carried out inside the condenser. The total heat transfer is called the Total Heat of Rejection (THR).

There are two major options used in India, air cooled condensers (cooling coil and fan) or water

cooled condenser, with cooling tower (see chapter 7.8). There are challenges with both—an air cooled chiller is not efficient in many climates in India due to the very high outdoor air temperature and water cooled chiller system with cooling tower consumes water.

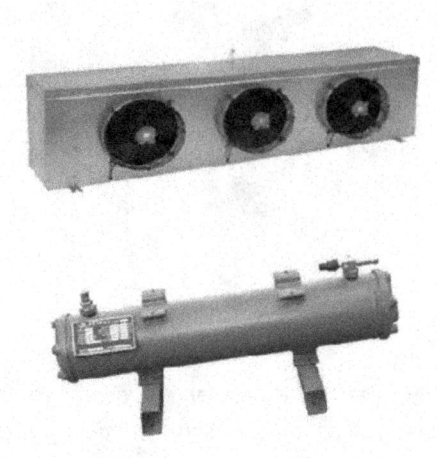

Figure 7.36. Air and water cooled condenser

The chilled water cycle comprises different kinds of valves, controls and safety devices, for which proper operation needs to be ensured during the start-up process. These include:

- High and low pressure limits to control chiller starts and stops;

- Freezing limit to prevent freezing in very low evaporating temperatures;

- Liquid receiver is a tank for liquid refrigerant;

- Filter drier to remove water and dirt from refrigerant;

- Refrigerant sight glass to see the total moisture content and the condition of refrigerant;

- Liquid line solenoid is a valve that controls the refrigerant flow;

- Hot gas bypass valve when load on an evaporator varies and operation of the air conditioning system is desired at lower than design conditions;

- Water makeup solenoid valve and reservoir level float switch to ensure the right water level in evaporator;

- Evaporator flow switch that prevents low flow through the evaporator. Very low or no flow can freeze the coolant inside the evaporator and destroy it.

Moisture is a problem in a refrigerant cycle. It can enter the system when the refrigeration system is being built up, it is opened for servicing or if leakage occurs on the suction side when it is under vacuum. Sometimes the system is filled with oil or refrigerant containing moisture. Moisture in the refrigeration system can cause blockage of the expansion device because of ice formation, corrosion of metal parts, chemical damage to the insulation or oil breakdown (acid formation).

The <u>filter drier</u> (Fig. 7.37.) is used to ensure that the system is internally clean and dry. During the operation, dirt and moisture must be collected and removed. This is performed by a filter drier containing a solid core consisting of Molecular Sieves, Silica gel or Activated aluminium oxide and a polyester mesh. Molecular Sieves retain water, whereas activated aluminium oxide retains water and acids.

A <u>sight glass</u> (Fig. 7.37.) with moisture indicator is normally installed after the filter drier. It indicates whether or not there is moisture in the refrigerant. Bubbles in the refrigerant indicate that pressure drop across the filter drier is too high, there is no sub-cooling or insufficient refrigerant has been used.

Thus, if indications of both the total moisture content in the refrigeration system and the condition of the refrigerant ahead of the expansion valve are required, a sight glass must be installed on both sides of the filter drier. Vertical mounting with downward flow means rapid evacuation/emptying of the refrigeration system. With vertical mounting and upward flow, evacuation/emptying takes longer because refrigerant must be evaporated out of the filter drier.

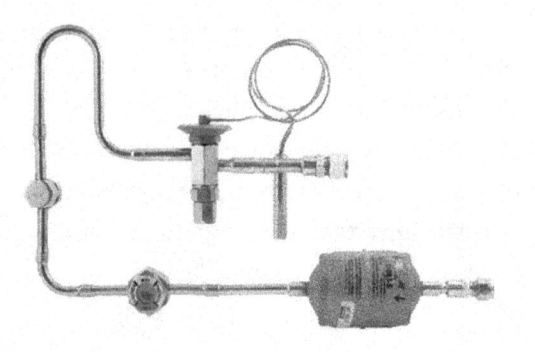

Figure 7.37. Chiller Expansion Valve, Sight Glass and Filter Drier.

Critical details for refrigerant piping is that they are made of the correct material specified in the construction documents (CDs), that all line sizes and pipe thicknesses are in accordance with the design, and that brazing piping joints are properly made and tested. The typical brazing processes is to use silver alloy solders and incorporate a dry nitrogen purge during brazing. Refrigerant piping shall be pressurized, evacuated, and tested for the sufficient time before insulation. Proper pipe saddles with correct distances shall be used during the installation. Also, appropriate expansion loops are required for the refrigeration piping.

Another critical detail on suction piping is that the return lines to the compressors slope downwards toward the rack systems for oil return and that suction risers, double pipe risers, and P traps are installed to facilitate oil return. It is also important that purge valves are installed at high points in condenser piping, that hot-gas defrost piping is sloped properly, and that liquid tees are taken from the bottom of the main liquid line.

A <u>refrigerant</u> is a substance or mixture, usually a fluid, used in a heat pump and refrigeration cycle. In most cycles it undergoes phase transitions from a liquid to a gas and back again. Refrigerants are divided into groups according to their chemical composition as follows:

- CFC = ChloroFluoroCarbons like R11 and R12 (banned due to the high ODP, high GWP);
- HCFC = HydroChloroFluoroCarbons like R22 and R123 (will be banned due to high ODP latest in 2030, high GWP);
- HFC = HydroFluoroCarbons like R32, R134a, R410A, R407C (ODP=0, GWP>1000 except R32 GWP=700);
- HC = HydroCarbon like R 290 (Propane) and R 600a (Isobutane) (ODP=0, GWP≈20);
- NH_3 = Ammonia R717 (ODP=0, GWP<0, low cost, but hazardous to humans);
- H_2O = Water R718 (ODP=0, GWP<1);
- CO_2 = Carbon Dioxide R744 (non-flammable, ODP=0, GWP=1, very low toxicity index and low cost but high operating pressure).

However, some of these chemical compounds are harmful to the environment (high Ozone Depletion

Table 7.3. Most common refrigerants used today and the future low GWP option. [39]

Sector	Current refrigerants used	ISHRAE assessment of low GWPoptions
Room	ACs HCFC-22, R-410A, HFC-32, HC-290	HC-290, HFC-32
Stand-alone units e.g. water coolers	HC-600a, HC-290, HCFC-22, HFC-134a, R-744	HC-600a, HC-290, R-744
Multi-s plit, VRF	HCFC-22, R-410A	none at the moment
Scroll chiller	HCFC-22, R-407C, R-410A	none at the moment
Screw chiller	HCFC-22, HFC-134a	none at the moment
Centrifugal chiller	HFC-134a, HCFC-123	none at the moment

Potential (ODP) and/or Global Warming Potential (GWP)), and therefore they should be replaced with more environmentally friendly alternatives. This is why the use of natural refrigerants like water, ammonia, carbon dioxide and hydrocarbons are becoming more popular. (Tab. 7.3.)

The chiller plant <u>control strategy</u> should be optimized for maximum efficiency at part load as well as full load operation. The size and consistency of loads will affect the optimum control sequences. With so many variables, no single control sequence will maximize the plant efficiency of all plants in all climates for all building types.

The typical design only optimizes a chiller plant for maximum load with the cooling towers and the condenser water pumps sized and set-up for maximum load operation and design conditions. While the chiller may have good part load performance, the rest of the plant is not effectively 're-tuned' to operate at part load, even though cooling tower fans do turn down based on a fixed condenser water temperature set point. Properly matching all the component operations under various loads and operating conditions will save significant amounts of energy.

Chillers are more efficient at higher leaving water temperatures so, in general, optimum efficiency is achieved when the chilled water supply temperature (CHWST) set point is as high as possible. Where all zones are controlled by the BMS system, the best reset strategy is based on valve position where the chilled water (CHW) set point is reset upwards until the valve controlling the coil that requires the coldest water is wide open. Another option is to reset the differential pressure set point alone or a combination of the temperature and pressure.

During operation of the chiller, if the condensing water temperature is below the design temperature, the capacity of the chiller is reduced because the mass flow capability of refrigerant is reduced. However, the efficiency of the chiller increases with lower condensing water temperature. A control strategy that reduces the condensing water temperature when the chiller is operating at part load will significantly improve the chiller's operating performance at that load and can lead to significant savings. This control strategy is defined as Condenser Water Temperature Reset.

With lower part load situations and lower outside conditions, the cooling tower fan will still be required to work harder than necessary to achieve the desired set point. But, if under these conditions the maximum fan speed is limited, the savings in cooling fan energy will outweigh the slightly increased energy usage of the chiller. A control strategy that is designed to achieve this balance is called Cooling Tower Temperature Relief.

During the design the Total Equivalent Warming Impact (TEWI) needs to be calculated. TEWI is a measure of the global warming impact of equipment based on the total related emissions of greenhouse gases during the operation of the equipment and the disposal of the operating fluids at the end-of-life. TEWI takes into account both direct fugitive emissions, and indirect emissions produced through the energy consumed in operating the equipment. [16]

TEWI is calculated as the sum of two parts, they are:

- Refrigerant released during the lifetime of the equipment, including unrecovered losses on final disposal;

- The impact of CO_2 emissions from fossil fuels used to generate energy to operate the equipment throughout its lifetime.

Total Equivalent Warming Impact can be calculated using the following formula:

TEWI = (GWP * m * L_{annual} * n) + (GWP * m *

(1 − $\alpha_{recovery}$)) + (E_{annual} * β * n)

where

- GWP = Global Warming Potential of refrigerant, relative to CO_2 (GWP CO_2 = 1);

- L_{annual} = Leakage rate p.a. (kg);

- n = System operating life (years);

- m = Refrigerant charge (kg);

- $\alpha_{recovery}$ = Recovery/recycling factor from 0 to 1;

- E_{annual} = Energy consumption per year (kWh/a);

- β = Indirect emission factor (kg CO_2 per kWh).

Another calculation that needs to made during the design is the total amount of refrigerants installed into a building in kg. This is a mandatory information that every building owner should know in many countries.

Fouling is a very undesirable phenomenon in any heat transfer unit. In mostheat exchangers, the fluid flowing is not completely free from dirt, oil, grease and chemical or organic deposits. In all cases, an unwanted coating can collect on the heat transfer surface, decreasing the heat transfer coefficient. The thermal efficiency of the heat exchanger will be reduced and the pressure drop

characteristics may change. The different types of fouling discussed are:

- Scaling;

- Particulate fouling;

- Biological growths;

- Corrosion.

Scaling is a type of fouling caused by inorganic salts in the water circuit. It increases the pressure drop and insulates the heat transfer surface, thus preventing efficientheat transfer. It occurs at high temperatures, or when there is low fluid velocity (laminar flow) and uneven distribution of the liquid along the passages and the heat transfer surface.

The likelihood of scaling increases with increased temperature, concentration and pH. Proper maintenance and treatment of the cooling water, e.g. pH treatment, greatly reduce the risk of scaling, especially in cooling towers.

Most scaling is due to either calcium carbonate (lime) or calcium sulphate (gypsum) precipitation. decreases with increasing temperature. The salts are deposited on the warm surface when the cold water makes contact with it. Pure calcium sulphate is very difficult to dissolve, which makes cleaning more complicated. In general, other types of scale are more easily removed.

The most important factors that influence scaling in the pipework are:

- Temperature;

- Turbulence;

- Velocity;

- Flow distribution;

- Surface finish;

- Composition and concentration of the salts in the water;

- Water hardness;

- pH

Scaling is more likely at a high pH, so a general approach to this problem is to keep the pH between 7 and 9. The risk of scaling generally increases with increasing water temperature.

The formation of calcium carbonate can be controlled by adding acids or specific chemicals tailored to inhibit the precipitation of the compound.

Fouling Factor (m², K/W) is the thermal resistance due to fouling accumulated on the water side or air side heat transfer surface. Chillers are tested and rated when the evaporator liquid-side, condenser liquid-side and air-side heat transfer surfaces are all clean. Therefore, the appropriate fouling factors shall be used when selecting the chiller and heat exchangers.

During the start-up it is important that the correct refrigerant is used and that there is no moisture or dirt in the refrigerant system.

Commissioning, tracking performance data, and paying attention to the details during start-up and shutdown, can raise the chiller performance while lowering expenses. Proper chiller initial and seasonal commissioning, seasonal start-up and shutdown procedures, along with a regular maintenance program, help to ensure that a chiller operates dependably and efficiently, resulting in reduced operating costs and improved occupant comfort throughout its life cycle.

The chiller commissioning process is about verifying the operating parameters identified in the specifications. Some of these parameters can be measured directly (e.g. temperatures, pressures, flow rates, power consumption), while others are calculated (e.g. refrigeration capacity, efficiency).

Baseline readings and data obtained during the commissioning process should be recorded for future use. This can be either done manually or, preferably, recorded through the chiller's control

Figure 7.38. Screw chillers usually used for continuous operation in commercial and industrial applications

panel or BMS via a history print. Understanding how the chiller was performing when it was in an as-new condition will provide a qualified technician with an optimal performance target throughout its life.

The other objectives of the commissioning process are identifying and correcting deficiencies in the system that could ultimately impact the reliability of the chiller and ensuring that the equipment is being operated in a safe manner. System deficiencies that could impact the reliability of a given piece of equipment include a water treatment program that is not in place or fully operational, as well as insufficient load that could cause the equipment to short cycle.

Before scheduling chiller commissioning, the technician who conducts the start-up should ensure all items on the delivery checklists pre-commissioning checklist are complete. Initial cleaning/flushing of the piping system should be performed in a manner that does not allow the cleaning solution or debris to enter the heat exchanger bundles. Failure to keep contaminants away from the tube surfaces may lead to accelerated failure of the chiller tube bundles.

Commissioning provides a great kick-start to ensuring that a chiller will perform optimally throughout its life. As with all mechanical equipment, normal wear and tear will occur over time, and predictive and/ or preventative maintenance must be performed on a regular basis to maintain chiller safety and performance. It is important to note that the parameters that were recorded at start-up were only a snapshot in time.

Once the chiller arrives on site, the following details need to be checked from the nameplate. Nameplate data must be in accordance with submittals as approved by the designer and commissioning agent:

- Manufacturer;
- Model no.;
- Serial no.;
- Evaporator model no. if not included in chiller no.;
- Condenser model no. if not included in chiller no.;
- Chiller type (centrifugal, screw, reciprocating, scroll);
- Condenser type (water or air cooled);

- Compressor motor:
- Volts;
- Locked Rotor Amps (LRA): current under starting conditions;
- Rated Load Amps (RLA): maximum current under any operating conditions;
- Refrigerant type
- Manufacturer's efficiency rating (kW/ton) in standard conditions and IEER.

Pre-commissioning includes among the other issues the installation verification. The following items needs to be validated before start-up and operational testing of chiller:

- Factory start-up sheet is completed and attached, for new construction this sheet must be completed before proceeding;
- Test and balance report is reviewed for chiller system flows;
- Chiller and environment are clean;
- There is adequate access for maintenance;
- No visible water or oil leaks;
- No unusual noise or vibration;
- Refrigerant fill is sufficient and refrigerant is clean from dirt and moisture;
- All chiller accessories are performing properly;
- Chilled water piping insulation is in good condition where visible;
- Pressure gauges & thermometer plugs installed where specified;
- Pressure gauges & thermometers installed where specified;
- Chilled water set point (panel readout) - acceptance ± 1 OC from design;
- Electrical current limit set point (panel readout) - acceptance ± 5% from design;
- Record and explain any diagnostic codes in control panel memory - acceptance: causes of all serious codes have been corrected;
- O&M manual on site.

Chiller start-up is typically carried out by the chiller manufacturer. Different type of chiller requires different activities, but the basic steps are common (Fig. 7.39.)

Controls calibration is an important part of the start-up. For thermostats, slowly adjust the set point until the controlled response begins (i.e. contact make or break). Note the set point when that occurs and the simultaneous measured value on a calibrated instrumentheld in close proximity to the sensing bulb. Check if sensor location is improper. Enter other chiller control points that are critical to the control sequence in the blank spaces for each chiller, as appropriate. Acceptance criteria are the desired temperature values ± 1 °C from measured values.

Chiller operation verification includes a series of field tests to verify that the chillers, as installed, operate as they were intended to operate by the manufacturer and designer. It is important to check if the field observation corresponds to the intended design operation.

All measured values must be within ±15% of the design values, unless otherwise noted under a specific test. Measured current (Ampere) must be less than the rated full load current. Measured voltage imbalance must be less than 2%.

Full load test is performed with the following tests and measurements by forcing the chiller to its maximum capacity. Any false loading should be done gradually to avoid overloading the systems. Loading can be done by some combination of increasing the building load (lowering cooling set points), heating the building, manipulating cross-over valves between the chilled water and condenser water piping, or manipulating the chilled water mixing valve on the chilled water return line.

Chiller operation test measurements shall include the following parameters:

- Chiller serial number;
- Chiller current: Rated Load Amps (RLA), design (A)
- Chiller current, measured (A)
- Voltage, phase to phase (V)
- Voltage imbalance, (max of a,b,c – avg)/avg) (%)
- Is voltage imbalance <2%
- Leaving chilled water (CHW) temperature, design (°C)
- Leaving CHW temperature, measured (°C)
- Entering CHW temperature, measured (°C)

- Entering CHW temperature, measured (°C)
- Delta (entering-leaving) CHW temperature, design (°C)
- Delta CHW temperature, measured (°C)
- Evaporator water flow rate, design (kg/s)
- Evaporator water flow rate, measured (kg/s)
- For water cooled condensers & cooling towers
- Ambient air temperature (°C)
- Entering condenser CHW temperature, design (°C)
- Entering CHW temperature, measured (°C)
- Leaving CHW temperature, design (°C)
- Leaving CHW temperature, measured (°C)
- Delta (leaving - entering) CHW temperature, design (°C)
- Delta CHW temperature, measured (°C)
- Condenser water flow rate, design (kg/s)
- Condenser water flow rate, measured (kg/s) For air cooled and evaporative condensers
- Ambient air temperature (°C)

During chiller system controls tests, the following tests that verify proper operation of the chiller and its auxiliaries under normal system control are carried out. If the actual control sequence differs from that implied by the tests, attach a description of the control sequence, the tests that were done to verify the sequence, and your conclusions. Use of data logging instrumentation is recommended to implement and document these tests, though visual observation is acceptable. Annotate any data and graphs so that it is clear what the data is proving. -

All data recorded during the start-up of the chiller shall be attached to the commissioning documents.

During the chiller system operation tests, the following data needs to be recorded:

- Chiller serial number
- Chiller appears to meet load (no complaints)
- Chillers operates without usual number of trips
- Chiller enabled under time- of- day control
- Outside air temperature lockout functions properly
- If CHW temperature is reset, what is the controlling independent variable
- CHW temperature follows reset schedule
- Chiller maintains CHW at set point ± 1 °C over a 2 hour operating period
- Remove the chiller load and verify the chiller and accessory shutdown sequence. Verified?
- Add load and verify the chiller and accessory start -up sequence. Verified?
- Shutdown and start-up sequences stage multiple chillers & accessories properly.

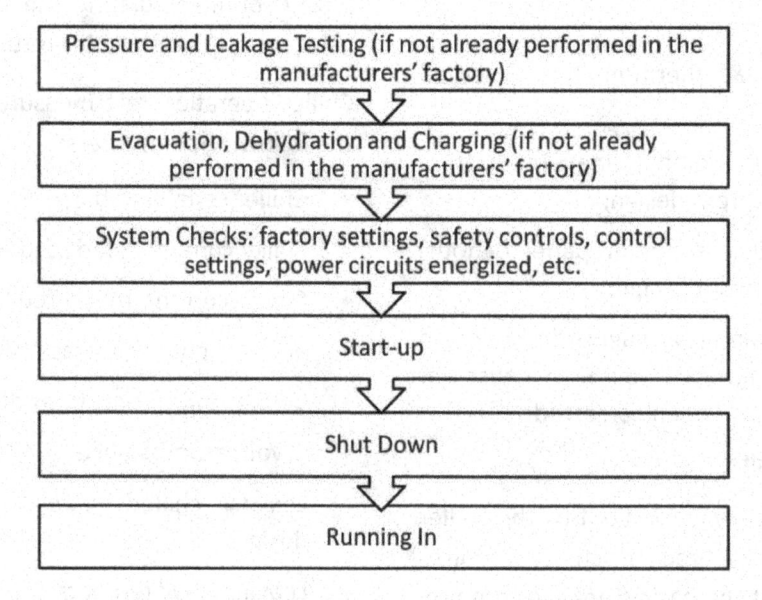

Figure 7.39. Chiller start-up procedure details are depending on the type of the compressor, but typically the following main steps are required.

Chiller Operation: Problems—Reasons—Solutions

Chiller operation is typically commissioned by the manufacturer. In case there are problems in chiller system operation, here are some typical reasons & solutions for them [26]:

- Chiller current, measured (A)

- Voltage, phase to phase (V)

- Voltage imbalance, (max of a,b,c - avg) / avg) (%)

- Is voltage imbalance < 2%?

- Leaving chilled water (CHW) temperature, design ($^{\circ}$C)

- Leaving CHW temperature, measured ($^{\circ}$C)

- Entering CHW temperature, design ($^{\circ}$C)

- Entering CHW temperature, measured ($^{\circ}$C)

- Delta (entering-leaving) CHW temperature, design ($^{\circ}$C)

- Delta CHW temperature, measured ($^{\circ}$C)

- Evaporator water flow rate, design (kg/s)

- Evaporator water flow rate, measured (kg/s)

For water cooled condensers & cooling towers

- Ambient air temperature ($^{\circ}$C)

- Entering condenser CHW temperature, design ($^{\circ}$C)

- Entering CHW temperature, measured ($^{\circ}$C)

- Leaving CHW temperature, design ($^{\circ}$C)

- Leaving CHW temperature, measured ($^{\circ}$C)

- Delta (leaving - entering) CHW temperature, design ($^{\circ}$C)

- Delta CHW temperature, measured ($^{\circ}$C)

- Condenser water flow rate, design (kg/s)

- Condenser water flow rate, measured (kg/s)

For air cooled and evaporative condensers

- Ambient air temperature ($^{\circ}$C)

During chiller system controls tests, the following tests that verify proper operation of the chiller and its auxiliaries under normal system control are carried out. If the actual control sequence differs from that implied by the tests, attach a description of the control sequence, the tests that were done to verify the sequence, and your conclusions. Use of data logging instrumentation is recommended to implement and document these tests, though visual observation is acceptable. Annotate any data and graphs so that it is clear what the data are proving.

All data recorded during the start-up of the chiller shall be attached to the commissioning documents.

During the chiller system operation tests, the following data needs to be recorded:

- Chiller serial number

- Chiller appears to meet load (no complaints)

- Chillers operates without usual number of trips

- Chiller enabled under time- of- day control

- Outside air temperature lockout functions properly

- If CHW temperature is reset, what is the controlling independent variable

- CHW temperature follows reset schedule

- Chiller maintains CHW at setpoint ± 1 $^{\circ}$C over a 2 hour operating period

- Remove the chiller load and verify the chiller and accessory shutdown sequence. Verified?

- Add load and verify the chiller and accessory start -up sequence. Verified?

- Shutdown and start-up sequences stage multiple chillers & accessories properly.

Room air temperature is too high:

- Pressure drop across evaporator is too high.

  - Replace expansion valve with a valve having external pressure equalization.

  - Reset superheat on expansion valve if necessary.

- Lack of sub-cooling ahead of expansion valve.

  - Check refrigerant sub-cooling ahead of expansion valve and establish greater sub-cooling.

- Pressure drop across expansion valve is less than the pressure drop the valve is sized for.

  - Check pressure drop across expansion valve.

  - Try replacement with larger orifice assembly

and/or valve.

- Reset superheat on expansion valve if necessary.

- Bulb is located after a heat exchanger or it is too close to large valves, flanges, etc.

  - Check bulb location.

  - Locate bulb away from large valves, flanges, etc.

- Expansion valve is blocked with ice, wax or other impurities.

  - Clean ice, wax or other impurities from the valve.

  - Check sight glass for colour change (green means too much moisture).

  - Replace filter drier if fitted.

  - Check oil in the refrigeration system.

  - Has the oil been changed or replenished?

  - Has the compressor been replaced?

  - Clean the filter.

- Expansion valve is too small.

  - Check refrigeration system capacity and compare with expansion valve capacity.

  - Replace with larger valve or orifice.

  - Reset superheat on expansion valve.

- Charge is lost from expansion valve.

  - Check expansion valve for loss of charge.

  - Replace expansion valve.

  - Reset superheat on expansion valve.

- Charge migration in expansion valve.

  - Check whether expansion valve charge is correct.

  - Identify and remove cause of charge migration.

  - Reset superheat on expansion valve if necessary.

- Expansion valve bulb not in good contact with suction line.

  - Ensure that bulb is secured on suction line.

  - Insulate bulb if necessary.

- Evaporator completely or partly iced up.

- De-ice evaporator if necessary.

Refrigeration system hunts:

- Expansion valve superheat set at too small a value.

  - Reset superheat on expansion valve.

- Expansion valve capacity too high.

  - Replace expansion valve or orifice with smaller size.

  - Reset superheat on expansion valve if necessary.

- Expansion valve bulb location inappropriate, e.g. on collection tube, riser after oil lock, or near large valves, flanges or similar.

  - Check bulb location.

  - Locate bulb so that it receives a reliable signal.

  - Ensure that bulb is secured on suction line.

  - Set superheat on expansion valve if necessary.

Suction pressure is too high:

- Liquid flow or expansion valve is too large or expansion valve setting incorrect.

  - Check refrigeration system capacity and compare with expansion valve capacity.

  - Replace with larger valve or orifice.

  - Reset superheat on expansion valve.

- Charge lost from expansion valve.

  - Check expansion valve for loss of charge.

  - Replace expansion valve.

  - Reset superheat on expansion valve.

- Charge migration in expansion valve.

  - Increase superheat on expansion valve.

  - Check expansion valve capacity in relation to evaporator duty.

  - Replace expansion valve or orifice with smaller size.

  - Reset superheat on expansion valve if necessary.

Suction pressure is too low:

- Pressure drop across evaporator too high.

  - Replace expansion valve with valve having external pressure equalization.

  - Reset superheat on expansion valve if

7.8. Cooling tower

A cooling tower is a heat rejection device which rejects waste heat to the atmosphere. Cooling towers typically use the evaporation of water to remove heat and cool the working fluid to near the wet-bulb air temperature. Condenser water can be cooled either in an open or closed circuit cooling tower. (Fig. 7.40. and 7.41.) In in an open circuit tower, the hot water from the condenser is directly spread to fill, where due to evaporation the water cools down. It is important to have a sufficient make-up water supply to compensate the amount of water that is evaporated.

Alternative to a cooling tower is the air cooler that typically contains a cooling coil and a fan. They rely solely on air to cool the condenser water or refrigerant to near the dry-bulb air temperature. A closed circuit cooling tower combines these two. The condenser water circulates in a closed loop, but separate water is used to create the evaporation in fill.

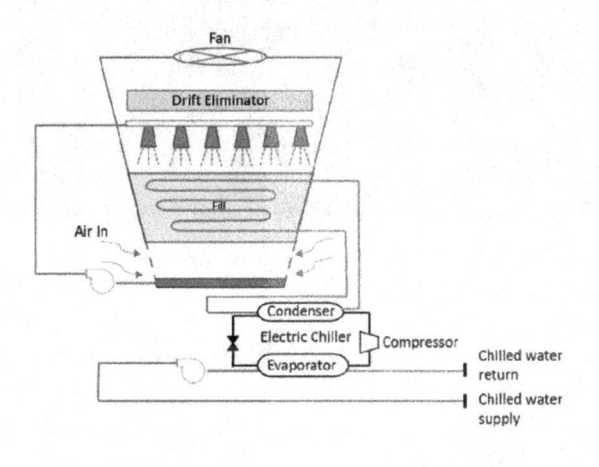

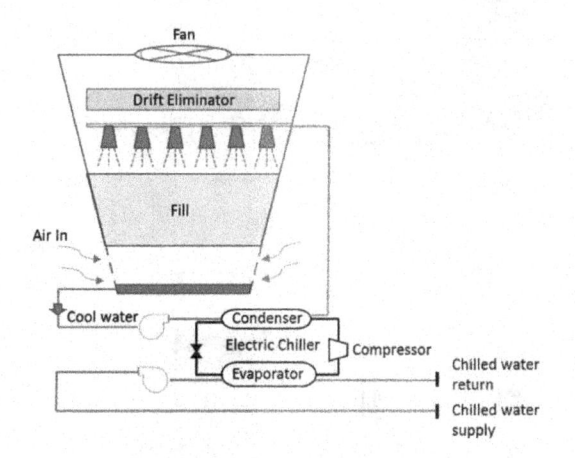

Figure 7.40. Chiller excess heat can be released either in open or closed circuit cooling tower.

Cooling tower requires the following checks during the pre-commissioning:

- The water-circulating system serving the cooling tower is thoroughly cleaned;
- That interior filling of cooling tower is clean and free of foreign materials such as scale, algae & tar;
- Fill is properly fixed;
- The cooling tower fans are free to rotate and the tower basin is clean (see also fan pre-commissioning criteria);
- The water circulating pumps are ready for test (see also pump pre-commissioning checks).
- There is no noise or vibration;
- Also check:
- Fixing of drift eliminator;
- Drive alignment and belt tension;
- Bearings & lubrication;
- Drainage & fall;
- Strainer cleanliness;
- Ball valve function;
- Tower water level;
- Water distribution;
- Water treatment equipment;
- Electrical supply connections;
- Earth bonding.
- During the cooling tower operational performance review, the following parameters (both designed and measured) are recorded:
- Heat rejection capacity (kW);
- Entering air dry bulb temperature (°C);
- Entering air wet bulb temperature (°C);
- Leaving air dry bulb temperature (°C);
- Leaving air wet bulb temperature (°C);
- Cooling water flow rate (kg/s)
- Cooling water entering temperature (°C);
- Cooling water leaving temperature (°C);
- Make up water flow rate (kg/s);
- Constant bleed water flow rate (kg/s);

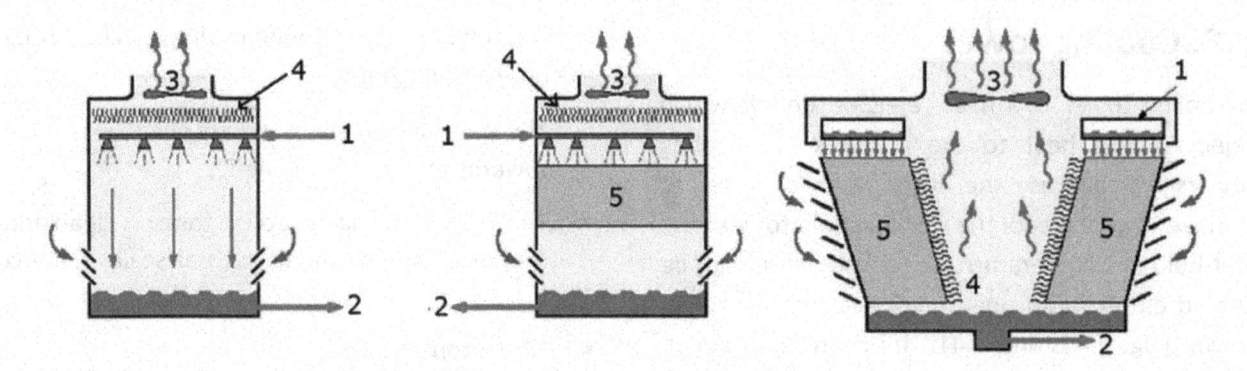

Figure 7.41. Different models of cooling towers from left: Induced draft counter-flow tower, induced draft counter-low tower with fill and induced draft cross-flow tower with fill. Each tower has the following components: 1) Hot water connection from chiller, 2) cold water connection to chiller, 3) fan, 4) drift eliminator and 5) fill.

- Fan type;

- Fan diameter (m);

- Fan air volume (m3/s);

- Fan power (kW);

- Fan pressure (Pa);

- Supply voltage (V);

- Motor starting current (A);

- Motor running current (A);

- Motor/fan speed (rpm);

- Phase motor insulation (RY/YB/BR) (M.ohm)

- Phase - earth motor insulation (RY/YB/BR) (M.ohm)

- Motor starter type.

All the cooling tower measurement results shall be recorded and attached to the commissioning data base.

7.9. Pump

Pumps are used in various applications in the building like distributing chilled water between a chiller and a cooling tower, distributing water into an air handling units' cooling coil and room cooling units, supplying water to a sprinkler system and as waste water pumps. (Fig. 7.42.)

Total head (Fig. 7.43.) and water flow rate are the main criteria that are used to select a pump for an application. It is the pressure between the inlet and outlet of the pump. The total head is the sum of static head and friction head. The discharge pressure of a pump depends on the pressure available on the suction side of the pump. Typically, the suction

head is less than the discharge head and therefore the static head (discharge head - suction head) is positive. Friction head is the amount of energy loss due to friction of the fluid moving through pipes and fittings.

Figure 7.42. Cooling water pumps are often installed parallel to each other.

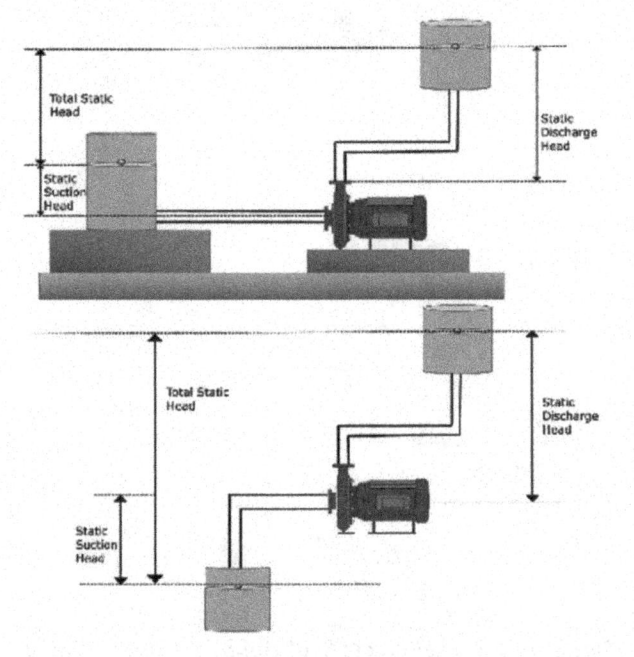

Figure 7.43. Total static head is the difference between discharge and suction head.

The head at any flow capacity is the sum of the static and the friction heads. (Fig. 7.44.) The static head does not vary with flow rate. It is only a function of the elevation or back pressure against which the pump is operating. The friction losses are related to the square of the flow, and represent the resistance to the flow caused by pipe and equipment friction.

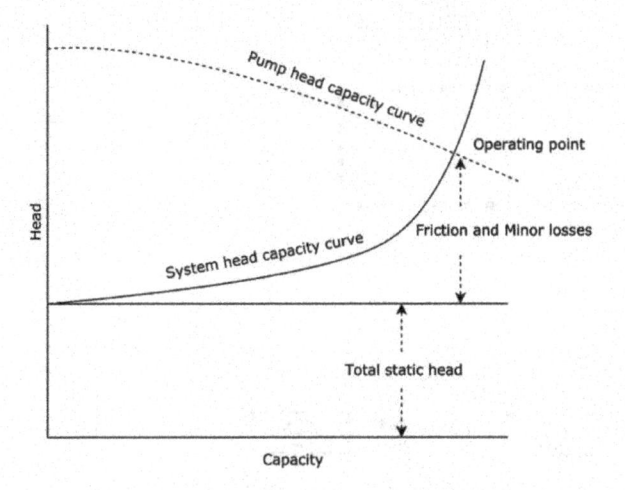

Figure 7.44. The characteristics of each unique system can be represented by a system head-capacity curve whereas a pump head-capacity curve describes the characteristics of the pump.

The head can be converted to a pressure using the following formulas (water pump):

- P (kPa) = 9.81 * head (m)
- P (bar) = 0.0981 * head (m)
- P (psi) = 0.43 * head (ft.)

Suction cavitation occurs when the pump suction is under a low-pressure/high-vacuum condition where the liquid turns into a vapor at the eye of the pump impeller. This vapor is carried over to the discharge side of the pump, where it no longer sees vacuum and is compressed back into a liquid by the discharge pressure.

Discharge cavitation occurs when the pump discharge pressure is extremely high, normally occurring in a pump that is running at less than 10% of its best efficiency point. The high discharge pressure causes the majority of the fluid to circulate inside the pump instead of being allowed to flow out the discharge.

Cavitation results in erosion e.g. at the impeller and high-intensity sound. Cavitation can be eliminated or minimized by removing restrictions at the pump suction. If the pump cavitates at high flows, a second pump should be started at a lower flow. If cavitation occurs at low flows, one might turn off the pump.

The efficiency of the chilled water system also depends on the pumping schemes. There are three basic schemes for delivering chilled water:

- Direct-Primary (constant volume) (Fig. 7.45.);
- Primary-Secondary (constant volume primary, variable volume secondary) (Fig. 7.46.);
- Advanced Direct-Primary (variable volume) (Fig. 7.46.).

Initially chilled water was delivered through a simple direct-primary, constant flow pumping scheme. This

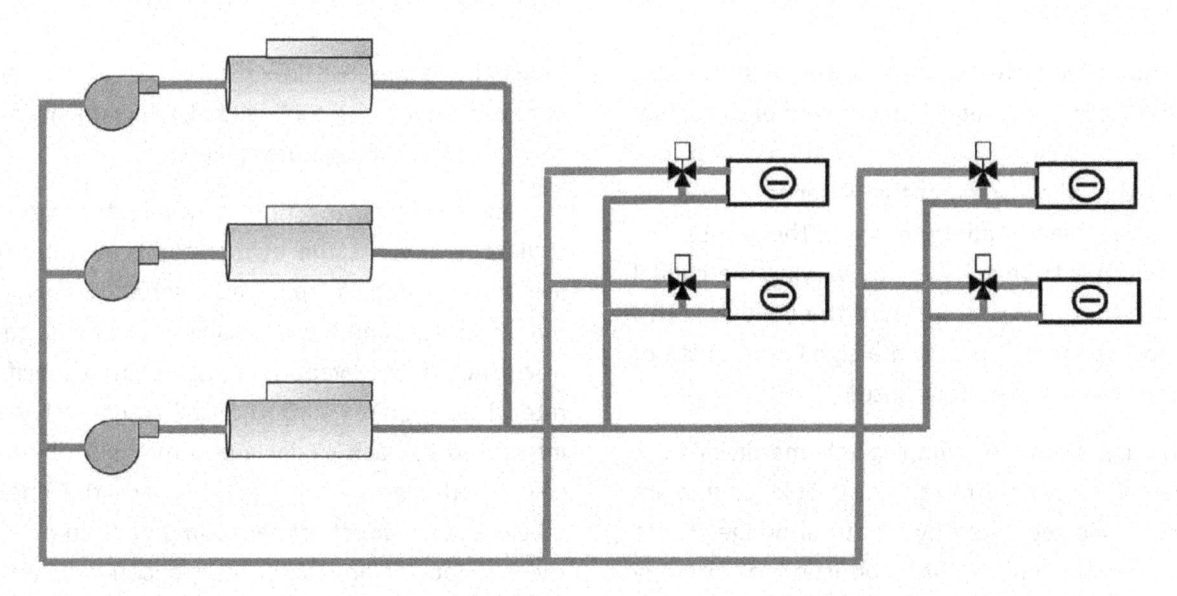

Figure 7.45. The traditional way to control water flow rates in a cooling system using constant volume pump and 3-way mixing valves in each cooling coil.

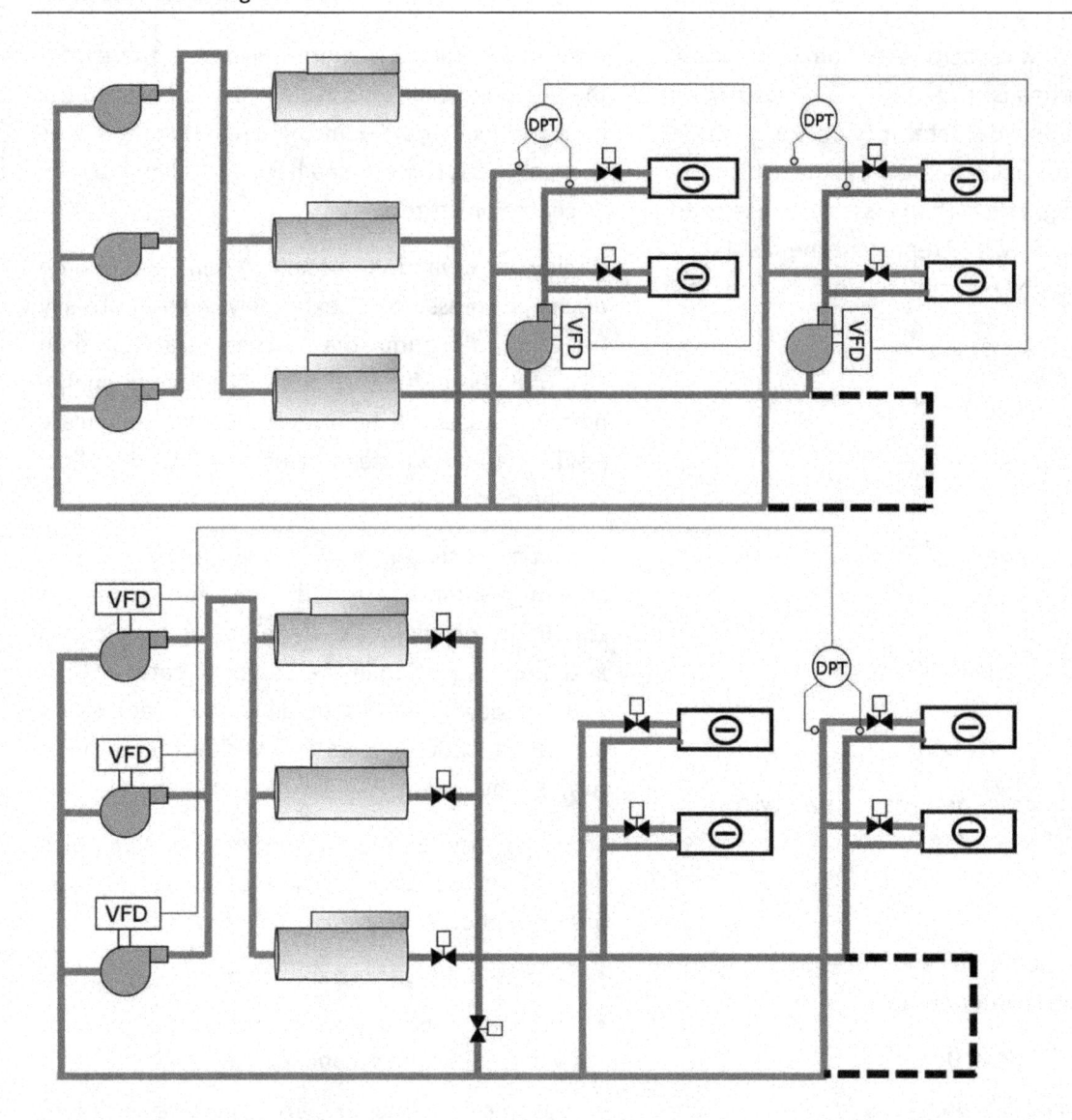

Figure 7.46. There are two optional ways to control water flow rates in a cooling system in an energy efficient manner: using constant flow primary pumps and variable flow secondary pumps with 2-way control valve in each AHU cooling coil (top) or using variable flow primary pumps, chiller control valves, minimum bypass valve in neutral bridge and 2-way control valves in AHU cooling coils (bottom). In both cases the pump is controlled by static pressure difference in the end of the pumping line.

method delivers chilled water to the end user at a constant flow rate that is independent of the actual load. Three-way control valves at the cooling coils are used to allow some of the water to bypass the coil during part load conditions. This varies the plant output to match load by varying the chilled water return temperature from the buildings. This method of control results in a significant waste of energy and loss of performance.

A primary-secondary pumping scheme divides the chilled water system into two distinct loops that are hydraulically separated by a neutral bridge. These loops are known as the production or primary loop and the distribution or secondary loop. The inherent separation of the primary and secondary

loops allows variable flow through the distribution system to match the cooling load, while maintaining constant flow through the chillers.

In newer chillers, the flow can vary through the chillers as long as the evaporator tube velocities are kept within a specified range (defined by manufacturer) and the allowable rate of change of flow through the evaporators is kept to a specified rate. This requires also the chiller control valve integrated to each chiller and a minimum by-pass valve in the neutral bridge. It allows the energy efficient use of direct-primary pumping by combining the two loops from the primary-secondary system into a single hydraulic loop and varying chilled water flow through the entire system.

The trouble-free pump operation is critical in terms of the total system operation. There are smart monitoring systems available to follow the operation of pumping station but the same diagnostics can be built into a BMS system. The following items shall be monitored: pump running or stopped, reason for stop (schedule/flow not required/tripped), static pressure difference in the system, valve positions. Based on this data, the system shall generate an alarm if necessary.

After the pump installation, the <u>pre-commissioning checks</u> shall be carried out as follows:

- Installation of chilled water pumps as per approved shop drawings;

- Installation of inertia base and vibration isolation springs;

- All nuts, bolts are tight and screwed on;

- Motors and pumps are aligned properly;

- Impeller is free to rotate when decoupled and should not make unusual noise when rotated by hand;

- Bearings are clean and lubricated and drive guards are fitted;

- Pressure test points are provided at the suction and discharge of pump;

- Flow measurement devices installation as per approved shop drawing;

- System has been flushed and cleaned in accordance with method statement;

- Pump strainer is cleaned after flushing and cleaning;

- Triple duty valve is provided at pump discharge;

- Pump drain is connected to main drain;

- Pump is insulated;

- Suction diffuser is installed at pump suction;

- VFD is installed and all controls connected.

Prior to pump start-up, the following checks need to be carried out:

- Check that all normally open isolating and regulating valves are fully open and that all normally closed valves are closed. In case of thermostatic valves, it is essential that provision for fully opening the valves is available. Most electric motorized valves have either provision for manual override of normal control using a switch on the main control box or a facility to position the valve seat mechanically.

- Check that the direction sign of all non-return valves is along the same discharge direction of associated pumps.

- Check that the horizontal or vertical alignment of all flexible joints is within the tolerances recommended by manufacturers' installation guideline.

- Open all control valves to full flow to heat exchangers of branch circuits.

- Fully open the return and close the flow valve on the pump. Close valves on standby pump. Closing the flow valve on the duty pump will limit the initial starting current, which is usually excessive at the first time a pump is running due to bearing stiffness.

The following measurements need to be carried out during the <u>operation test</u> and respectively data shall be recorded to the commissioning documents:

- Water flow rate;

- Total head;

- Mark the measured operation point to the pump performance curve;

- Measure the current and voltage.

7.10. Flushing, venting, insulating and balancing of pipework

There are several pipe works in the building. In this chapter the focus is mainly on the chilled water pipework. The potable water, grey water and sprinkler pipe works have their own specific criteria.

Chilled water pipework is typically made either from copper or plastic pipe. Big main pipes are sometimes made using steel (iron) pipes. Multi-layer Cross Linked Polyethene (PEX) or Polyethylene of Raised Temperature Resistance (PE-RT) plastic pipes are used especially in radiant cooling applications.

When plastic or other non-oxygen tight materials (e.g. rubber) are used, it is important to ensure that air does not enter into the closed hydraulic system. In the plastic pipe this is ensured by using the multilayer structure, where a thin layer of aluminium is added into a pipe structure. This needs to be ensured also with flexible and other pipe connections.

Air creates corrosion and air cushions in a hydraulic system. Considerably larger amounts of sludge are formed, and this may lead to clogged valves and pipes, pump problems and circulation disturbances. This does not happen immediately, but problems are eventually very likely to occur. Air gets into the system with make-up water or is trapped in the system after the initial filling. Air can also get diffused through the pipe and other components (all metals are airtight). Also, air ingress occurs due to negative pressure. In case a non-oxygen tight system is used, then corrosion resistant components or anti-corrosion additives must be used.

After the installation of pipework, <u>pre-commissioning checks</u> are conducted:

- Piping is complete;

- As-built shop drawings are submitted;

- Isolation, control and other valves are installed as required;

- Operation of valves is verified;

- Air-vents (manual or automatic) are installed as specified: pipes should be installed so that they do not leave any single 'air pockets', and a venting valve should always be installed at the highest point of the vertical main pipes in the shaft;

- Flexible connectors are installed as specified;

- Verify that piping has been labelled and valves identified as specified.

Fabricated welded joints in piping system require various types of quality assurance tests such as simple visual scanning and inspection, non-destructive tests, and occasional mechanical test, e.g. hardness and corrosion resistance test:

- Visual inspection: using lenses, CCTV and fibre optic cameras;

- Dye penetration testing: staining surfaces with fluorescent liquid which show flaws under ultraviolet lighting;

- Magnetic particle testing: applying magnetic particles to magnetized materials so that they line up along the cracks and defects (often combined with dye testing);

- Radiographic testing: using x-ray to find internal faults;

- Ultrasonic testing: firing high frequency sound pulses into materials and analyzing the differences in reflected signals.

After the pipework pre-commissioning checks are done, it is important to flush the pipework in order to remove all construction debris, dirt and particulate matter. (Fig. 7.47.) This should be carried out such that no sensitive parts (e.g. chillers, heat exchangers or small diameter radiant cooling pipes) of the hydraulic system suffer damage during the flushing. The bypass of these components needs to be planned during the design or it needs to be provided by contractor before flushing.

Fill-in/flushing water needs to be tested (total viable counts (TVC), pseudomonas and sulphate reducing bacteria absent) before flushing to analyse its cleanliness for hydraulic use. Also after the flushing, the water sample needs to be analysed to ensure successful flushing. After filling the system, the flushing and water cleaning should be done in 48 hours in order to avoid biofilm development or consequent bacteria-induced problems.

Figure 7.47. Water is full of debris and other dirt at the beginning of flushing.

Before filling up, all shut-off and control valves must be in the fully open position. Pumps should not be running during the filling-up (static filling). Continuous venting is necessary during the filling and it is recommended to have both manual and automatic venting systems installed. (Fig. 7.48.)

Pump should only be started when filling is complete. To remove all air from the system, the major part (>75%) of the system should be closed so that the water can circulate fast enough. When each section is full, it should be closed, and the same procedure repeated for the rest of the system.

Figure 7.48. Automatic and manual venting valves.

During the flushing several pumps may need to operate at the same time depending on the application and size of the system: system duty pump alone, together with standby pump or both together with temporary flushing pump. The primary pump strainer basket shall not be less than 3 mm. when flushing starts.

Flushing velocity is dependent on the pipe size varying between 0.96 m/s (190 fpm) to 1.26 m/s (250 fpm) with pipe sizes of 15 mm to 150 mm respectively. Pressure drop across pumps need to be monitored so that the manufacturer's maximum pressure is not exceeded and pump cavitation problems are avoided.

Drainage of flushing water needs to be planned. Adequate foul drains are required near the flushing points. In case specific cleaning chemicals are used during the flushing, this needs to be taken into account when planning the drainage of flushing water. Also a temporary flushing tank may be required.

Dynamic flushing is recommended in several steps in order to keep the water velocity high enough. First the primary main-ring circuit shall be flushed, thereafter the main pipes and finally horizontal mains in each floor. After each round of flushing the strainer basket needs to be changed. Each loop needs to be flushed until the water is clean and

the drainer basket does not collect sediment. The flushing of horizontal main pipes shall start from the top floor through intermediate floors and finish in the lowest floor.

After these pre-flushing, the full system flushing shall be carried out to dilute and replenish the system water from soluble iron level.

Dynamic flushing shall be done based on BSRIA Application Guide AG 1/2001 and its revision 2004. [18]

After filling, flushing and venting a system, the <u>pressure test</u> needs to be carried out based on the following principles:

- Water piping shall be hydrostatically tested to a pressure to 1-1.5 times design pressure;

- Test shall be performed for at least 24 hours with fluctuation in pressure;

- Pressure testing shall be performed prior to installation of thermal insulation. All joints must be fully exposed to environment;

- Underground and other piping that will be subsequently get covered after construction shall not be covered without prior approval of successful testing.

Pipe <u>insulation</u> can be done only after pressure testing of system. Chilled water pipes operate at below-ambient temperatures and therefore the water vapour condenses on the pipe surface. Moisture creates many different types of problems like corrosion, so preventing the formation of condensation on pipework is usually considered important. Pipe insulation is also important to reduce the energy loss in the pipework.

Pipe insulation is the only way to prevent condensation on the pipe surface. When the insulation surface temperature is above the dew point temperature of the air, condensation will not occur. The insulation shall also create a water-vapour barrier that prevents the latter from passing through the insulation to the pipe surface, where it would condensate.

Therefore, the following items in insulation need to be checked:

- Insulation material is sufficient for its use: thermal conductivity, water-vapour resistance, surface emissivity and insulation protection from ambient conditions like solar, temperature and rain;

- Thickness of insulation is as per design documents;

- There are no damages in the insulation material surface;

- Insulation material is well connected to the pipe surface so that there are no air pockets between insulation and pipe.

The aim of water balancing is to apply a standard method of adjustment to water flow rates throughout the system to meet the particular requirement of the design. Water balancing can be done using proportional balancing method in a similar way than proportional ductwork balancing (chapter 7.2):

- Keep all the balancing, control and isolating valves at fully open position;

- Measure the initial water flow rates across each balancing valve;

- Record the water flow rate and compare it with the design flow rate (ratio);

- Find out the index valve (lowest ratio of flow) in each branch;

- Keep the index valve at fully open position;

- Throttle first the furthest valve (reference) to the same ratio and after that every other valve in branch starting from the lowest ratio;

- Monitor the index valve percentage after throttling each valve as it increases gradually;

- By the time the last valve is completed, the flow is balanced at all valves including the index valve;

- Record the readings in all valves and attach it to the commissioning documents.

In case pressure independent balancing and control (PIBC) valves are used, proportional balancing is not required, though each valve is set to correct water flow rate as per manufacturer's instructions.

7.11. Balancing and control valves

Balancing valves maintain the designed water flow condition in the designed operation point so that control valves may function properly and provide correct flows to a system under the changing conditions. (Fig. 7.49.) The control valve regulates the water flow rate typically based on the air temperature either after the air handling unit or in the space. There are two options for balancing and control of water flow rates:

- Balancing valve with adjustable K-value and separate control valve (2-way or 3-way valve depending on the application);

- Pressure independent balancing and control (PIBC) valve.

HVAC system flows are dynamic, changing during the day and night. Because of external heat gains and changes in the building occupancy, the demand for cooling and heating varies not only throughout the day, but also by building sector. Proper hydronic balancing is the key to making an HVAC system perform properly and cost effectively.

The balancing valve has typically an adjustable pressure loss. This ensures that the pressure difference in the system can be adjusted so that each pipe branch has the same relative pressure loss and therefore correct water flow rate in design conditions. The designer would have calculated the kv-values for each valve. However, there may be changes in pipework made after the design and therefore pre-calculated kv-values may not give correct water flow rates but pipework balancing is required (see chapter 7.10). In most cases the balancing valve also has measurement taps for water flow measurement.

Pressure independent balancing and control (PIBC) valves compensate for pressure variations in the pipework, performing a continual balancing function to maintain system performance at varying loads. Predictable flow eliminates over-pumping and equal percentage flow leads to system controllability. Constant flow performance significantly reduces actuator movement, providing less hunting and wear on the valve assembly. Valves are selected based on flow rate and no kv-calculations or pipework balancing are needed. Each valve is set to its operation point and water flow rate is measured. In this case separate control valve is not needed.

Figure 7.49. Example of balancing valve with adjustable kv-value (left) and pressure independent balancing and control valve without and with actuator (middle & right). Both valves have the water measurement taps integrated.

Flow accuracy of the PICB valve is ±10%. However, actuator hysteresis and installation can have an effect on measured accuracy of the assembly (actuator/valve) in the field. The accuracy of the PICCV assembly can be improved in the application.

The control valve controls the water flow rate based on the temperature measurement. The actuator can be connected directly to the local controller or via BMS system. Two-way control valve opens and closes typically based on the 0-10 V control signal. Valve operation can be proportional 0-100% or on-off depending on the application. Three-way control valve keeps the water flow rate constant in the main circuit by adjusting the water flow rate running via heat exchanger and bypass. If the control valve has the adjustable kv-value, it can be used for balancing as well.

Items to <u>check before starting</u> the flow verification procedure:

- Verify that water is clean; the system is purged of air and filled to proper pressure.

- Verify that each valve has high enough pressure to operate, e.g. pressure independent control and balancing (PICB) valve has at least 35 kPa (5 psi) but less than 350 kPa (50 psi) pressure difference across the valve.

- Verify proper pump operation.

- Verify that supply water temperature is at design temperature.

- Ensure that strainers are clean.

- Check that all manual shutoff valves are open.

- Check that all bypass valves are closed.

- In case the system has both the PICB and manual balancing valves, all manual balancing valves must be set to their operating (see chapter 7.10) point before PICB valve operation are verified.

- Pressure and water flow rate measurements needs to be recorded in every control valve and results are attached to the commissioning documents.

7.12. Unitary products

A unitary system typically combines heating, cooling, and fan sections with all necessary controls in one. (Fig. 7.50) They require only few assemblies, mainly the refrigeration pipework. As most of the functions are built in the factory, the commissioning requirements are less. The contractor shall carry out visual checks and preliminary checks as recommended by the manufacturer before start-up of the equipment.

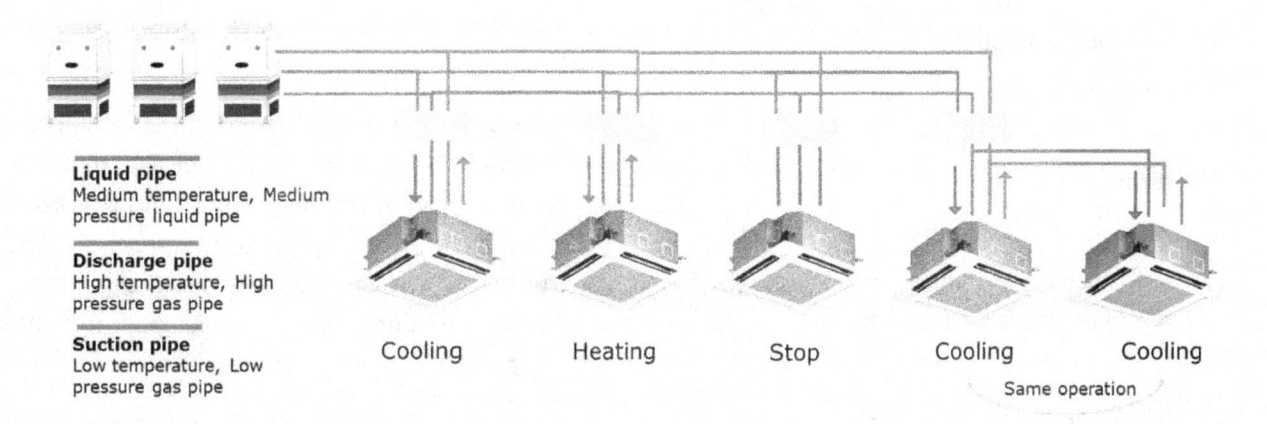

Figure 7.50. The refrigerant flow is varied using either an inverter controlled variable speed compressor or multiple compressors of varying capacity in response to changes in the cooling or heating requirement within the air conditioned space.

- Pre-commissioning tests of split units and VRF-units:

- Check that the indoor and outdoor units are installed according to the supplier's instructions and recommendations including maximum pipe run, etc.;

- Check that the safety devices and necessary operational control are provided as per requirements of general specifications;

- Check that the pipe works of correct pipe sizes are installed;

- Inspect all welding and conduct the pressure test before closing walls or ceilings;

- Pressure and leak tests of refrigerating piping systems shall be carried out on completion of the system but before pipework is insulated;

- Check that the units are correctly wired up;

- Ensure that all the connection nuts are tightened;

- Ensure that condensing water pipes are correctly installed and clean;

- Make sure that all isolating valves are open where appropriate.

During the start-up:

- Carry out the water drainage test;

- Trial run the units only after water drainage test;

- Measure the following and record the results:

  - Ambient and indoor air temperature;

  - Starting and running current and voltage;

  - High and low pressure cut out;

  - Compressor suction and discharge pressures;

  - Evaporator entering and leaving coil temperature both WB & DB;

  - Condenser entering and leaving coil temperature, both WB & DB;

  - Air flow rate;

  - Noise level measurement;

  - Other measurements recommended in manufacturer's instructions;

- Record also:

  - Indoor and outdoor unit serial numbers;

  - Fuse size and type;

  - Refrigerant type;

- Refrigerant pipe length;

- Factory refrigerant charge, refrigerant added on site and total refrigerant charge in kg;

- Test the control system according the manufacturer's instructions;

- Ensure that system evacuation, dehydration and charging with refrigerant has been carried out, vacuum achieved and is done by skilled and experienced personnel.

7.13. Radiant cooling

Radiant cooling system family consists of various types of solutions:

- Chilled ceiling panels;

- Embedded ceiling cooling;

- Thermally activated building systems (TABS);

- Floor cooling or heating.

Chilled ceiling panels are metal or gypsum ceiling tiles with integrated copper or plastic pipes. They are usually suspended from the concrete slab-like normal suspended ceiling. Typically, there is an aluminium diffuser or graphite filling behind the surface panel to accumulate the heat from the entire panel surface. Lower thermal mass compared to other radiant cooling solutions means they cannot be used as a thermal storage. (Fig. 7.51.)

However, the temperature control in panels reacts more quickly to changes in indoor air temperature and is therefore better suited to spaces that have a big variance in cooling loads. Perforated panels also offer better acoustical properties in case acoustic material has been used above panels. Panels are connected typically with flexible hoses to header pipes.

Embedded ceiling cooling system consists of plastic pipes in the screed below slab. Typically, there is a layer of insulation between screed and concrete slab to prevent thermal energy to spread to the concrete slab. In this solution pipes are typically very small (< 6 mm/0.2 in) to enable application of the screed. [16]

Each pipe loop is a single pipe whose ends are connected to a manifold. Typically, 4-8 pipe loops can be connected to a one manifold. Each manifold has the shut-off valves for each loop, in some cases

Figure 7.51. Different radiant cooling solutions: chilled ceiling panels (top left), embedded ceiling cooling elements (top right), thermally activated building systems (TABS) pipes installed into a slab before slab casting (bottom left) and floor cooling pipes installed before casting a screed.

the control valves, venting valve and bypass between inlet and return side in the manifold for flushing purpose. As the pipe is very small and cannot be changed after screed is applied, it is very important that no dirt gets into the pipe loops.

In Thermally Activated Building Systems (TABS) the plastic pipes are cast inside the concrete slab. Pipes are laid and pressure tested before casting. In this case the entire concrete slab performs as a thermal energy storage. It is very important to protect the pipes through the construction process. No drilling of slab shall be allowed before careful planning as the pipes cannot be fixed in case they get damaged. In this case the pipes are typically bigger (15-20 mm/0.6-0.8 in) than in embedded ceiling system. Each pipe loop is connected using the manifold as in an embedded cooling system. [17]

In floor cooling or heating system plastic pipes are embedded into the screed above the slab. Typically, there is insulation between screed and concrete slab to prevent energy loss. There are several installation options depending on the manufacturer and the pipes are installed into either a new construction or a refurbished building. Manifold is used to connect each pipe loop to a main cold water circuit.

All radiant cooling systems are designed to operate dry (no condensation) and therefore the inlet water temperature has to be above the room dew point temperature. Typically, this means temperatures between 15 - 20 °C (59 - 68 °F). This needs to be taken into account during the fill up of the system and start-up of system operation. In case the air inside the building is humid during the fill up, water temperature has to be even warmer (above dew point). (Fig. 7.52.)

Condensation (sweating) happens on any surface that is colder than the dew point temperature of room air. If condensation occurs, either the surface temperature needs to be increased or the moisture content (absolute humidity) of air needs to be reduced by increasing the dehumidification in AHU cooling coil or using separate room air dehumidifiers.

The control sequence of the entire system must be checked in BMS to ensure that there is no risk of condensation in any operation condition and, if condensation takes place, there must be sufficient safety features to close the water circulation and prevent water damages for property.

Plastic pipes shall be oxygen diffusion resistant multilayer (plastic-metal-plastic) PEX or PE-RT pipes to avoid problems with corrosion and air in the closed hydraulic system (noise & poor heat transfer). All inlet pipes need to be insulated in order to avoid condensation and reduce heat losses.

Before start-up of the chilled water system, the following <u>pre-commissioning checks</u> need to be carried out:

- The chilled water system cannot be operated before the dry system operation is ensured. Operate the primary air handling unit for dehumidification of air inside the building before commissioning the chilled water system. Close all operable windows, and ensure building exit doors are sealed to assist in the envelope dehumidification.

- Ensure that the main pipes have been flushed and pressure-tested prior to being connected to the radiant panels directly or to manifolds. In case they are already connected during the flushing, use the bypass line to prevent dirty water from entering into manifolds, panels or radiant pipework.

- Confirm that all air has been removed from the distribution pipes. Deliver excess water by increasing the pump flow, or by closing other zones to assist in the removal of air from the system.

During the <u>start-up process</u> of the cooling system:

- The required water flow rates are adjusted during commissioning.

- The chilled water system is commissioned as described in chapters 7.10 and 7.11.

- Once the designed dew point of air inside the building has been reached, slowly lower the secondary chilled water temperature to the scheduled design value. Note that dehumidifying the building envelope may require several days, or up to a week initially, to completely dehumidify the space.

- The commissioning of the chilled water circulation systems is carried out by balancing the water flow rates using balancing valves and ensuring that all the shut-off valves are open.

- The water flow rate is controlled by a control valve, which is typically an on-off or modulating valve.

- Balancing valves can be replaced with control valves with adjustable kv-value.

- In case pressure-independent control valves are used, follow the manufacturer's instruction to set the correct water flow rate.

After the water system start-up:

- The radiant system behaviour must be verified using a thermal camera. Surface temperature of ceiling/floor in each pipe loop needs to measured and recorded on the commissioning documents.

- Record the balancing and control valve set values in the commissioning documentation.

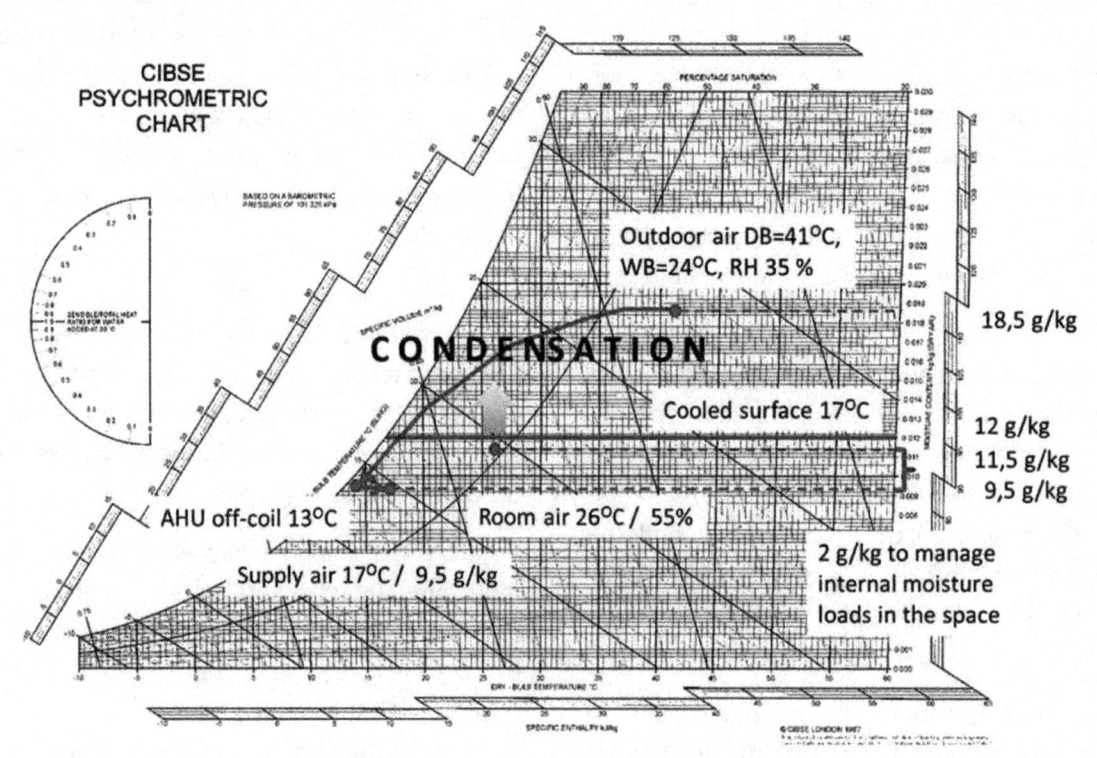

Figure 7.52. Supply air needs to be dry enough to manage the internal moisture loads in order to ensure dry operation of radiant cooling system.

7.14. Chilled beams

Chilled beam systems are used for cooling and ventilating spaces, where good indoor environmental quality and individual space control are appreciated. (Fig. 7.55.) Chilled beam systems make use of a dedicated outdoor air systems. This system is to be applied primarily in spaces where internal humidity loads are moderate. They can also be used for heating.

Active chilled beams (Fig. 7.53 and 7.54) are connected to both the ventilation supply air ductwork, and the chilled water system. When desired, hot water can be used in this system for heating purposes. The main air-handling unit supplies primary air into the various rooms through the chilled beam. Primary air supply induces room air to be recirculated through the heat exchanger of the chilled beam. In order to cool or heat the room, either cold (15-18 °C/59-64 °F) or warm (30-45 °C/86-113 °F) water is cycled through the heat exchanger. Recirculated room air and the primary air are mixed prior to diffusion in the space. (Fig. 7.56.) [68]

Room temperature is controlled by the water flow rate through the heat exchanger.

Passive chilled beams comprise of a heat exchanger for cooling and heating. The operation is based on natural convection. The primary air is supplied to the space using separate diffusers either in the ceiling or wall, or alternatively through the raised floor.

The chilled beam system is also always a dry system and condensation should not occur in any surface of a beam (cooling coil, pipe or duct). Read more about condensation management in chapter 7.13.

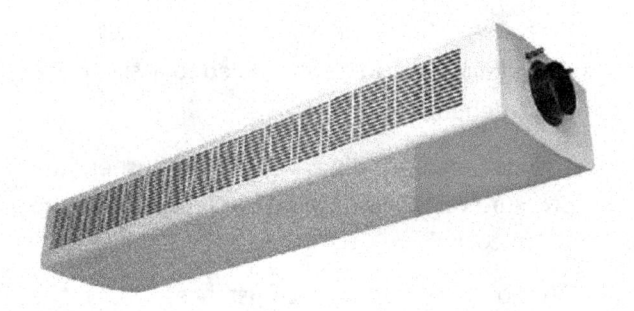

Figure 7.53. Active chilled beam for exposed installation.

Before the start-up of chilled water and ventilation systems, the following <u>pre-commissioning checks</u> need to be made:

- Ensure that the right model is installed into the right place (total & active length/nozzle type);

- Confirm primary air ducting is free of dirt and debris to prevent beam nozzle clogging.

- Ensure that all ducts and access ports are affixed to the duct to achieve specified duct leakage rates.

- Slit protective film at the active beam discharge to allow primary air to enter the space. Do not remove the protective film, until the workspace is in an 'as-new' condition.

- Ensure a clean environment within which the active beams will operate (i.e. no gypsum dust or other construction contamination).

- Remove protective film from surfaces of active beams only after all interiors are cleaned and free from dust.

- Do not operate the active beams for temporary cooling. In case this has been done, make sure that cooling coils are properly cleaned off dirt.

- The chilled water system cannot be operated before the dry coil operation can be ensured. Commission and operate the primary air handling unit for dehumidification of air inside the building before commissioning the chilled

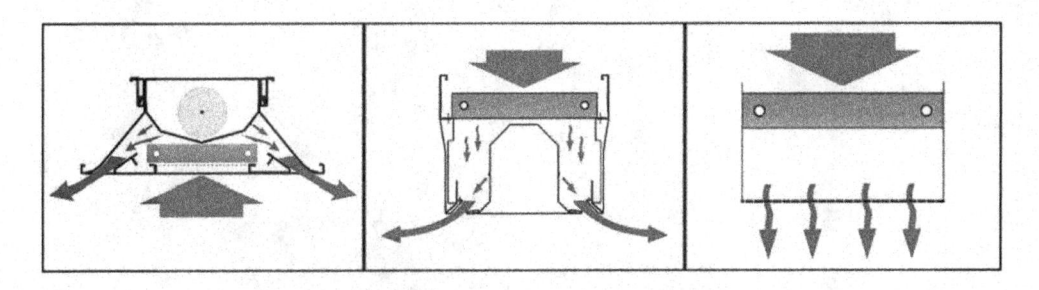

Figure 7.54. Operation principle of closed active, open active and passive chilled beams. In active chilled beams the primary air supply induces room air through a heat exchanger whereas in passive chilled beam (right) the operation is based on natural convection. [68]

water system. Close all operable windows, and ensure building exit doors are sealed to assist in the envelope dehumidification.

- Ensure that main pipes have been flushed and pressure tested prior to being connected to the beam coils.

- Do not, under any circumstances, flush the pipes through the beam's heat exchanger coils.

- Confirm that all air has been removed from the distribution pipes. Deliver excess water by increasing the pump flow, or by closing other zones to assist in the removal of air from the system.

- Confirm secondary water conditions regularly to ensure that is it properly filtered, and appropriately inhibited.

During the start-up process of ventilation and cooling system:

- The required air and water flow rates are to be adjusted during commissioning. Ventilation system and chilled water system are commissioned as described in chapters 7.1, 7.2, 7.10 and 7.11.

- Balance supply air ducting in each zone. Chilled beam operation is very sensitive to primary air volume.

- The airflow rates are typically adjusted with an individual VCD or constant flow damper.

- Measuring the airflow rate by using a chamber pressure measurement in the beam is recommended. This gives the most accurate measurement result due to the high pressure level (50-150 Pa). In other methods, e.g. pitot-tube measurement, the pressure level is much lower.

- Once the designed dew point of air inside the building has been reached, slowly lower the secondary chilled water temperature to the scheduled design value. Note that it may require several days, or up to a week initially, to completely dehumidify the envelope and the space.

- The commissioning of the chilled and hot water circulation systems is carried out by balancing the water flow rates using balancing valves and ensuring that all the shut-off valves are open.

- The chilled beam water flow rate is controlled by a control valve, which is typically an on-off or modulating valve.

- Balancing valves can be replaced with control valves with adjustable kv-value.

- In case pressure-independent control valves are used, follow the manufacturer's instruction to set the correct water flow rate.

After the air flow and water system balancing:

- Check the chilled beam function with a thermal camera. Direct the camera towards the chilled beam supply air slot after maximum cooling capacity is set on the room controller.

- Temporarily lowering the chilled water set point to approximately 10°C (50°C) during commissioning will highlight any malfunction of the system (too low water flow rate, shut off valves closed, etc.).

- Record the balancing and control valve set values, VCD set values in zones or each beam and measured chilled beam chamber pressures in the commissioning documentation.

Figure 7.55. Active (left) and passive (right) chilled beam installation.

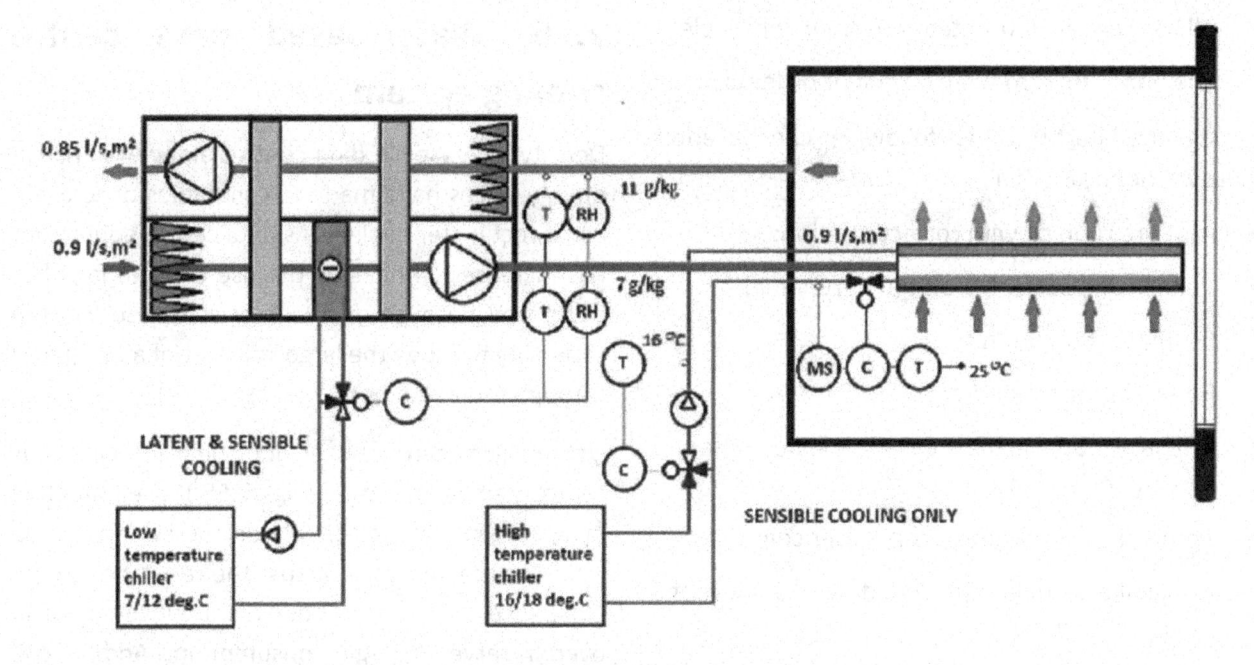

Figure 7.56. Typical operation schematic of active chilled beam and DOAS systems.

7.15. Precision air conditioning

Precision air conditioning systems (Fig. 7.57.) are designed to meet the needs of heavy electronic loads. Electronic loads generate relatively dryer heat as compared to typical comfort cooling conditions. Since the latent loads are almost negligible in spaces like data centres, precision air conditioning systems are specifically designed for catering to sensible loads only.

Before start-up, the following pre-commissioning checks need to be done:

- Equipment are located to provide adequate service access;

- Proper electrical supply voltage is available;

- Fused disconnect switch sized in accordance with the requirements;

- Proper size fuses in the disconnected switch;

- Proper size wiring to the unit;

- All wiring inside the unit is connected tightly to the terminals;

- Thermostat is provided and wired in properly;

- All wiring checked for proper hook up;

- Fan turns freely by hand;

- Condenser and evaporator fan set screws are tightened on the flat of the shaft;

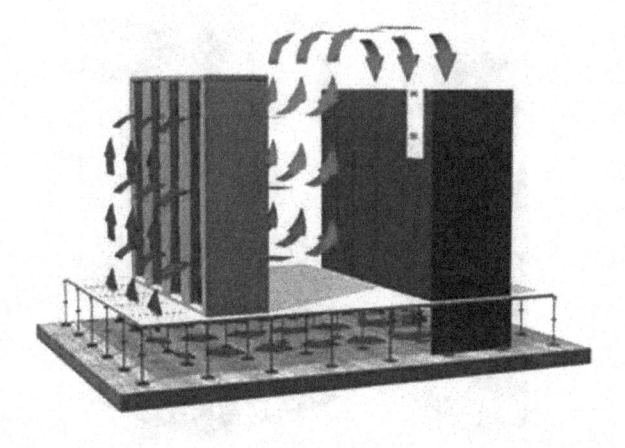

Figure 7.57. Precision air conditioning unit supplies air to the raised floor plenum from where it is supplied to the cold aisle. Cooled air flows via data racks to the hot aisle from where it is sucked to the air intake of precession air conditioning unit.

- All bearings are lubricated, wherever applicable;
- The unit is free from refrigerant or oil leaks.

During the <u>start-up</u>, the following checks and measurements are done:

- Fans are running with correct rotation;
- Voltage and current at unit (unit running);
- Supply air temperature;
- Return air temperature;
- Pressure loss across each module: filters, cooling coil and fan;
- Ambient temperature at condenser coil;
- All capillary tubes are tied down to prevent excess vibration;
- The unit is free from rattles.

The ventilation system and chilled water system are commissioned as described in chapters 7.1, 7.2, 7.10 and 7.11. After the air flow and water system balancing, check the chilled beam function with a thermal camera. Direct the camera towards the chilled beam supply air slot after maximum cooling capacity is set on the room controller. Temporarily lowering the chilled water set point to approximately 10°C (50 °F) during commissioning will highlight any malfunction of system (too low water flow rate, shut off valves closed, etc.). But before doing so, make sure that the air in the room is dry enough to avoid condensation.

7.16. Water-based data centre cooling system

Density increases in data centres along with rise in energy prices have made IT companies once again use direct water cooling. As data centre equipment has increased in density, the use of large fans has decreased because of a lack of space; so has the capability to move the large volumes of air required through the equipment.

Data centres can use rear door heat exchanges as self-contained systems. (Fig. 7.58.) Their limiting factor is that they must rely on the server fans to push the hot server air across the resistance of the cooling coil. Server fan power is a critical factor in overall server energy consumption. Adding coil resistance to the server fans increases their power draw.

Active rear doors (Fig. 7.58.) comprise a chilled water coil, variable speed (EC) fans, water control valve and PLC controls. They are dynamic and able to provide the exact cooling capacity, airflow and water flow for any changing heat load up to their rated capacity.

In-row coolers are situated between racks and contain a cooling water coil and fan(s). They draw air from the hot aisle and discharge cooler air to the cold aisle. The server fans draw this cooler air across the servers.

Figure 7.58. Passive and active data centre rear door coolers (left) as well as the in-row cooler (IRC) (right) are used in data centres.

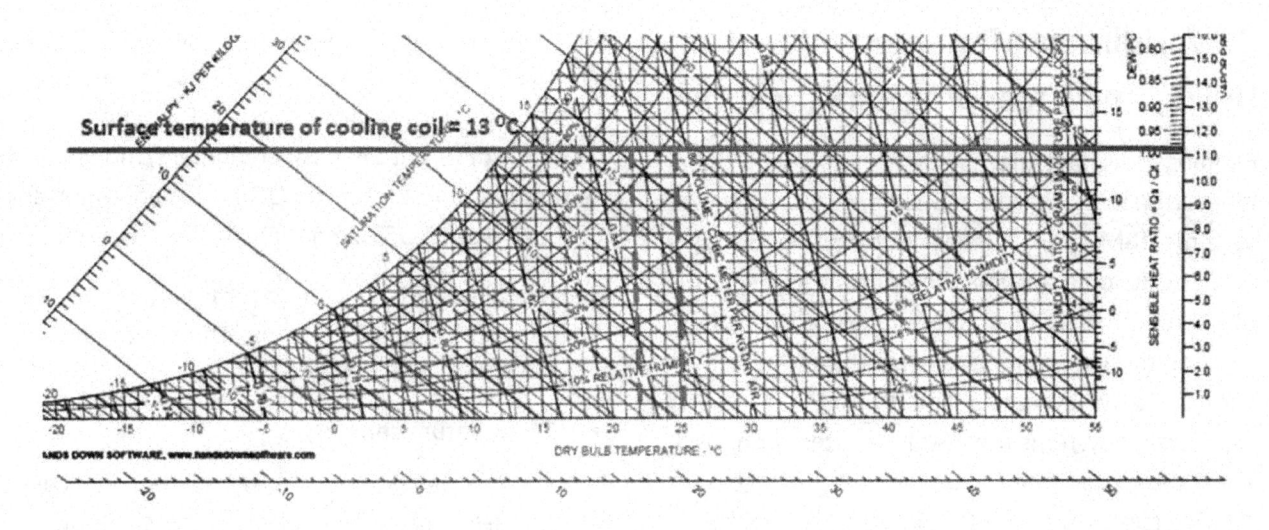

Figure 7.59. Condensation occurs in the cooling coil surface when the humidity ratio of a room air is above the dew point conditions. As an example, if the surface temperature of the coil is 13 °C (55 °F), the room air dew point therefore has to be maximum 12.9 °C (54 °F) to avoid condensation. This is equal to the humidity ratio level of 11 g/kg (68 grains/pound). In case the room air temperature is 22 °C, (72 °F) the relative humidity of the room air can be a maximum of 58% and respectively with the room air temperature of 25 °C, (77 °F) it can be maximum 47%.

Even though data centre mainly produces sensible heat only, condensation (sweating of cooling coil) often occurs during the start-up of the system. (Fig. 7.59.) This happens because there is still significant moisture in the air in the construction site during the start-up. Condensation can also occur during the operation in data centres where dehumidified air is not supplied and the room is not in the positive pressure. Therefore, the condensation collection system (trays, pipes and pumps, if necessary) is required to collect the condensing water. Positive pressure can only be created by supplying outside air into a space.

The methods to ensure that the absolute (and relative) humidity is low enough inside the data centre are:

- Supply dehumidified air into a space;

- Install room air dehumidifiers into a space.

- Measure temperature and relative humidity and ensure that room conditions are always below dew point conditions of supply water.

Before start-up, the following pre-commissioning checks need to be done:

- A chilled water system cannot be operated before the dry system operation is ensured.

- Ensure that the main pipes have been flushed and pressure-tested prior to being connected to the cooling panels. In case they are already connected during the flushing, use the bypass line to prevent dirty water from entering into panels.

- Confirm that all air has been removed from the distribution pipes.

- Ensure proper electrical supply voltage is available (in case the fan is integrated).

- Make sure that a thermostat is provided and wired in properly.

During the start-up, the following checks and measurement are done:

- The required water flow rates are adjusted during commissioning. The chilled water system is commissioned as described in chapters 7.10 and 7.11.

- Once the designed dew point of air inside the building has been reached, slowly lower the secondary chilled water temperature to the scheduled design value.

- The commissioning of the chilled water circulation systems is carried out by balancing the water flow rates using balancing valves and ensuring that all the shut-off valves are open.

- Air temperature before and after data rack is measured or operation is verified using the thermal camera.

- Record the balancing and control valve set values into commissioning documentation.

7.17. Building Management System (BMS) and measurement sensors

Building Management System (BMS) and measurement sensors Building Management Systems (BMS) or Building Automation Systems (BAS) or Building Automation Controls and Systems (BACS) controls building operations like ventilating, air-conditioning, heating, electrical power, lighting and security systems. Often fire alarm and suppression systems are also integrated with BMS, even they have their independent management system too. (Fig. 7.60.) BMS has the ability to monitor and control systems to maintain the indoor environment in acceptable limits, improve system performance, and conserve energy and water.

Commissioning a BMS takes place in the final phase of a building project when the project is often already late. Delays and shortage of skilled staff that have already occurred in a project tend to pile on the pressure. Some of the pressure can be eased by completing work such as configuring software, off site. By extending commissioning after the building is occupied, the control system may not perform as well as it should before commissioning is fully completed.

Building systems and equipment operate based on a designed Sequence of Operations (SOO) typically developed by the design engineer (specification) and modified during the submittal process by the contractors. The SOO is reviewed by the Cx Provider who utilizes the SOO to develop the Functional Performance Test (FPT) procedures including different systems, equipment, modes, and sequences of operations. SOO is also an important part of Systems O&M manual.

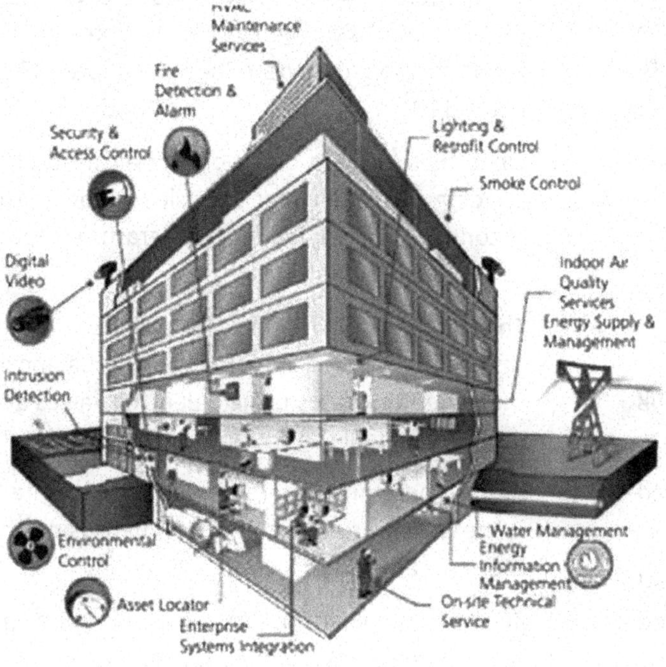

Sensors (Fig. 7.61.) are the critical part of the BMS system and building operation. The typical sensors used in HVAC applications are:

- Temperature sensors:
  - Thermocouples (robust but not sensitive, used in high temperature applications);
  - Resistance Temperature Detectors (RDS);
  - Thermistors.
- Pressure sensors:
  - Capacitive pressure sensors (for measuring differential pressure, small, can operate in high temperature, both static and dynamic pressure measurements);
  - Inductive pressure sensor (for low pressure applications, both static and dynamic pressure measurements);
  - Piezo-electric pressure sensors (for widely fluctuating pressures).
- Flow rate sensors:
  - Pitot tubes (used inside duct applications, highly robust, used e.g. in VAV dampers);
  - Orifice plate (inside duct or pipe, measuring the pressure difference across the plate, robust, often used with dirty liquids);
    - Venturi meters (same measurement principle than with orifice plate, large, expensive);
    - Hot wire anemometer (sensitive enough to detect low air velocity/flow rate, for free air movement as well as ducts);
    - Turbine flow meters (used liquid-in-pipe applications, for clean water only, used in BTU meters).

Figure 7.60. Building Management System controls all operations in the building. Some major or critical components have their own control system like fire alarm panel and chiller panel

- Humidity sensors:
  - Hygrometers (non-linear and prone to drift);
  - Psychrometers (distilled, water-wetted wick around temperature sensor to measure wet-bulb temperature, high velocity required, not suitable for HVAC control);
  - Resistive and capacitive sensors (electronic sensor with sensitivity of 2%);
  - Dew-point sensor (indicates the conditions where the condensation starts).
- Comfort sensors:
  - PMV-sensor (measures air temperature, mean radiant temperature and air speed);
- Indoor air quality sensors:
  - CO_2-sensors (duct and wall mounted);
  - RSPM-sensors (PM2.5, PM10);
  - TVOC-sensors;
  - Formaldehyde sensors;
  - Carbon monoxide sensors.
- Occupancy sensors:
  - Ultrasonic motion sensors (fills the room with continuous high frequency sound waves, high sensitivity to small movements, vulnerable to false triggering);
  - Infrared motion sensor (sensing moving infrared heat emitting objects, good sensitivity in short distances, better immunity against false triggering).

When the sensors are selected and purchased, the following criteria should be considered:

- Performance based factors:
  - Range;
  - Accuracy;
  - Resolution;
  - Repeatability (ability to continuously reproduce the same output);
  - Sensitivity (smallest detectable range);
  - Drift (degree to which the sensor fails to give a consistent performance);
  - Linearity (closeness to linear proportionality between output and measured value across the range);
  - Response time (time to elapse to get an output value response);
  - Noise of interference (unwanted signals);
- Practical and economic factors:
  - Cost;
  - Maintenance;
  - Compatibility (interoperability and interchangeability with other components and standards);
  - Environment (ability to withstand harsh or hazardous environments).

The Cx provider works closely with the controls contractor to verify the control programming and identify corrective issues during reviews and the FPT. During the FPT, deficiencies in the system operation or controls can be found. Every point and sequence is typically not required to be tested during the FPT.

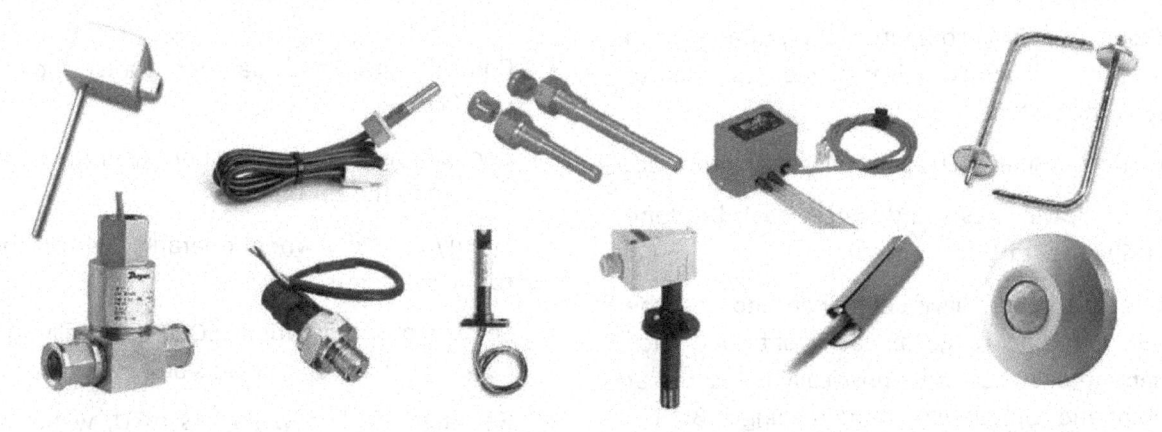

Figure 7.61. From the top left: duct air temperature sensor, water temperature sensor, thermal well for temperature sensor, static duct pressure sensor, static pressure measurement tubes for duct installation; from the bottom left: water pressure differential transmitter, water pressure sensor, hot wire anemometer for duct installation, duct humidity sensor, dew-point sensor and occupancy sensor.

Contractors generally hold the responsibility of testing and verifying every point and sequence. The Cx provider performs a test over a sample of the items after the contractors have tested, repaired and verified all issues.

The following <u>pre-commissioning checks</u> need to be carried out:

- Check installation of all instruments and measurement sensors on field (location, installation position, wiring, etc.).

- Verify that the controller power supply is from the emergency power supply, if applicable.

- Make sure that water or refrigerant temperature sensors are installed using proper thermal wells.

- Ensure that static pressure measurement sensor in the ductwork and the differential pressure measurement sensor in the pipework are correctly located and that the data cable is properly installed between the sensor and the actuator.

- Test end-to-end continuity of all cables (part of electrical commissioning).

- Ensure that power supply is available to Motor Control Centres (MCC), controllers and firefighting panels and that they are correctly installed.

- Ensure that all pressure and temperature gauges are installed.

- Ensure that all labels and nameplates are in place to recognize each sensor or component and pair them with the measurement or operation points in BMS.

- Ensure that all meters (energy, BTU, gas, water) are properly installed and connected.

- Ensure that BMS computers and printers are in place and all software is installed and ready to start the testing.

- Verify that spare I/O capacity has been provided.

- The following tests and checks shall be done during the start-up:

- After controllers have been energized, test the full operation range of each controller. Each parameter needs to be physically measured at point and compared with the reading in BMS.

- Check that the set points for each controller are correctly set in BMS and adjust PID controller actions.

- Check analogue inputs and outputs with 0, 50 and 100 per cent of span.

- Check digital inputs using jumper wire and outputs using ohmmeter to test for contact making or breaking.

- Calibrate pressure transmitters at 0%, 50% and 100% span and set differential pressure flow transmitters for 0% and 100% values with 3-point calibration accomplished at 50%, 90% and 100% span.

- Calibrate pressure switches to make or break.

- Check fan and pump activation output and control.

- Check the chilled water control valves' positions and feedback;

- Check the chilled water pressure sensor at supply and return header;

- Check the water flow switch operation in the chiller;

- Check the readings of VFD in pumps and fans;

- Check energy recovery wheel operation sequences;

- Test software and hardware interlocks.

The FPT plan specifies all the tests carried out to ensure proper system operation. Finally, <u>test each system for operation</u>, record readings and attach them to the commissioning documents.

- Temperature, RH, CO_2 and other sensor readings;

- Fan, chiller and pump start-up and shutdown sequences;

- Fan and pump speed and status indicator readings;

- Control valve and damper status indicator readings;

- Control valve and damper operation with respect to the set point;

- Energy recovery wheel operation with respect to the set points;

- Sequence of Operation (SOO) especially in air handling units, fans, chillers and pumps;

- Readings of all flow meters (BTU, water, gas, electricity);

- Functional tests including the airflow or water flow fail, alarm reset and manual override.

Table 7. Examples of BMS Data Points and Field Devices [35]

Description	Function			Field Device
	Status	Alarm	Control	
Water Chilling Units				
Chiller on/off	X		X	Chiller panel
Chiller water temperature reset	X		X	Chiller panel
Chiller run status	X			Chiller panel
Chiller trip fault		X		Chiller panel
Chiller water temperature in/out	X			Chiller panel
Chiller flow switch	X			Flow switch
Chiller out butterfly valves/ status	X		X	Motorised valve
Chiller current monitoring	X			Current transduced
Condenser water temperature in/out	X			Chiller panel
Condenser flow switches	X			Flow switch
Condenser out valve/ status	X		X	Motorised valve
Heat/ cool change-over valve/ status	X		X	Motorised valve
Chilled water flow rate	X			Paddle type flow meter
Thermal storage motorised valve	X		X	Motorised valve
Chilled Water Pumps (secondary/ primary/ condenser)				
Pump on/off	X		X	Pump panel / control ler
Pump status	X	X		Pressure switch
Pump auto/manual status	X		X	Electric panel
Static pressure difference in the pipework	X			Static pressure sensor/ pipe
VFD frequency control	X		X	VFD panel
Cooling Towers				
Cooling tower on/off	X		X	Local panel
Cooling tower status	X	X		Current relay
Cooling tower sump low water	X	X	X	Level indicator
Cooling tower valves in	X		X	Motorised valves
Cooling tower valves out	X		X	Motorised valves
Cooling tower auto/ manual status	X		X	Local panel
Indoor and Outdoor Air Conditions				
Air temperature	X			Temperature sensor
Relative humidity	X			Relative humidity sensor
CO_2-level	X			COr sensor
RSPM	X			RSPM sensor

Description	Function			Field Device
	Status	Alarm	Control	
Air Handling Units				
AHU on/off	X		X	AHU panel
Air flow status	X			Air flow DP switch
AHU filter pressure difference	X	X		AHU panel
Chilled water control valve modulation	X		X	Motorised 2- or 3-way valve
Static pressure in the duct	X			Static pressure sensor/ duct
VFD frequency control	X		X	VFD panel
Supply air temperature	X			Temperature sensor
Return air temperature	X			Temperature sensor
Fresh air damper status	X			Potential free contact from damper actuator
Fire alarm signal	X	X		Potential free contact from Fire Alarm Control Panel FACP
AHU auto/ manual status	X		X	Electric panel
Energy Recovery Units				
Energy recovery unit on/ off	X		X	Electrical panel
Energy recovery unit status	X			Current relay
VFD frequency control	X		X	VFD panel
Motorised damper status	X		X	Damper actuator
Ventilation (exhaust, smoke extract, pressurization)				
Fan on/ off	X		X	Fan panel
Fan status	X			Ventilation panel
Pressurization status	X	X		E.g. lift panel
VAV-damper / status	X		X	Motorised damper
Fire Fighting Systems				
Jockey, hydrant, diesel and sprinkler pump status	X			Pump panel
Sprinkler line pressure	X	X		Pressure sensor
Hydrant line pressure	X	X		Pressure transducer
Fire pump status: FAS interfacing	X			Pump panel
Sprinkler flow switch: FAS interfacing	X			Pressure sensor
Smoke exhaust fan on/off	X		X	Fan panel
Staircase pressurization fan on/off	X		X	Fan panel
Smoke exhaust fan status	X	X		Ventilation panel
Staircase pressurisation status	X	X		Pressure sensor
Fire and smoke damper status	X			Damper actuator

7.18. BTU meter

BTU meters measure the energy content of liquid flow in British Thermal Units (BTU), a basic measure of thermal energy. BTU meters (Fig. 7.62.) are used in chilled water systems for both commercial and industrial and office buildings. These meters are used to bill users for energy usage.

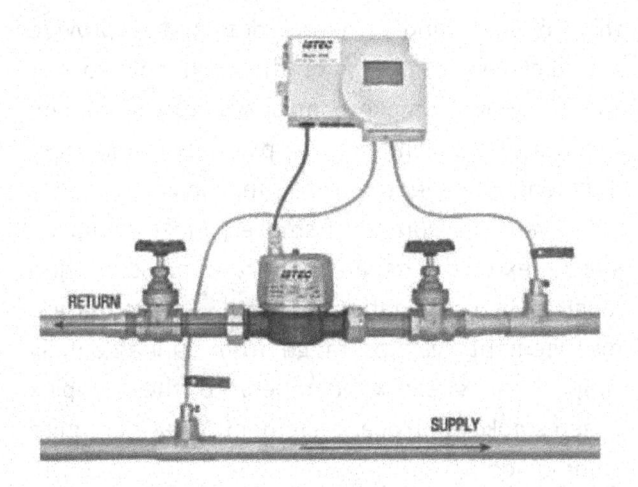

Figure 7.62. A BTU meter measures the water flow rate and the temperature difference between supply and return water pipe.

Commissioning and calibration of BTU metering devices are the processes that contribute to the proper installation and provision of adjustments necessary to generate accurate meter data. The accuracy and reliability of the data obtained from the meters largely depends on whether they are properly installed or calibrated. (Fig. 7.63.)

The commissioning process involves verifying the mechanical installation, measuring flow and temperature signals and then comparing these measurements to the specified installation and operating parameters listed on the certificate of calibration provided with the meter. The data collected during this initial commissioning process will then serve as the baseline data for periodic revalidation of the meter operation.

Basic BTU flow meter configurations are:

- Full bore/inline (volume or velocity);
- Insertion (velocity);
- Clamp on (velocity).

Important points to review during <u>pre-commissioning</u> are:

- Flow meter location is as prescribed in the plans;
- There is adequate straight pipe run before and after meter. Compare actual straight pipe run upstream and downstream of the flow meter location to the recommended distances identified in the flow meter installation manual.
- Confirm that the meter is tagged for the pipe diameter and the material it is installed in and that this information corresponds to that listed in the BTU meter certificate of calibration.
- Confirm that the temperature sensors are properly installed.
- Verify the type of fluid used in the piping system and confirm that the fluid specified on the BTU meter certificate of calibration matched the fluid flowing in the piping system.
- Confirm correct AC voltage is available at the power supply input terminals.

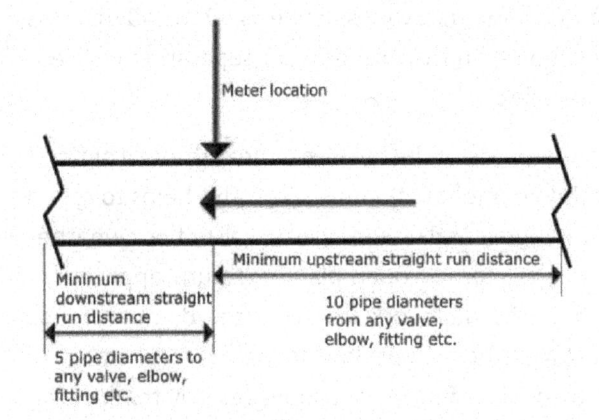

Figure 7.63. Correct location of BTU meter.

7.19. Fire and smoke control systems

Fire and smoke safety systems are integrated into the building to prevent fire damages, allow occupants to evacuate safely and ensure easier access for firefighters to enter the fire zone in the building. The system consists of several components (Fig. 7.64.):

- Fire dampers in supply ducts;
- Fire dampers in return air openings or ducts

- Sprinkler system;
- Staircase pressurization fans (see chapter 7.20);
- Smoke extract fans and dampers;
- Fire and smoke alarm system.

A fire damper is designed to close automatically upon detection of heat inside the duct or it can be connected to the fire and smoke alarm system. In case of fire, it interrupts the supply airflow, resists the passage of flame further inside the ductwork and maintains the integrity of the fire rated separation. Its primary function is to prevent the passage of flame from one side of a fire-rated separation to the other.

Fire dampers are operated by a fusible link or can be actuator operated (actuator spring closes the damper if there is no power). Fire dampers are equipped with a fusible link or temperature sensor that is rated for 74 °C (165 °F) up to 140 °C (284 °F) depending on the application. Fire dampers are selected based on their fire resistance (heat & flames) time that varies between 30 to 240 minutes depending on the fire rating of separation where it is installed.

Fire dampers are installed in or near the wall or floor, at the point of duct penetration. This helps to retain the integrity and fire rating of a wall or floor whether it is a ducted or open-plenum return application. Should the ductwork fall away, the damper needs to stay in the wall or floor to maintain the integrity of the wall or floor. One should actually think of the fire damper as part of the wall system itself. If a fire damper is installed further away from the wall, the distance between the wall and the damper has to be fire insulated with the same rating of the damper itself. It is also very important to check that the screed filling in the space between the actual wall and damper has same fire rating in case the damper is installed into a wall structure.

A smoke control system is a system that is used to limit the migration of smoke within a building due to fire. There are several methods to limit this migration and some are designed to provide a tenable environment for the occupants to exit the building. A smoke control system can include fire and smoke dampers that prevents smoke from migrating outside the zone, mechanical systems to prevent the spread of smoke (smoke dampers, smoke extract fans and staircase pressurization fans) or a combination of both. Smoke hazard management systems range from fundamentally simple fire & smoke dampers to the complex zoned smoke control arrangement. They all require commissioning.

Smoke dampers can be opened or closed from a remote fire command station, if required. An integrated fire and smoke damper's primary function is to prevent the passage of smoke through the ventilation system from one side of a fire-rated separation to the other. Or smoke extract dampers are opened in the zone where fire is simultaneously with smoke extract fan and closed in all other areas.

Smoke dampers are operated by either a factory-installed electric or a pneumatic actuator. They are controlled by smoke detectors and/or fire alarms. There are two general applications: 'passive smoke control system' in which they close upon detection of smoke and prevent the circulation of air and smoke

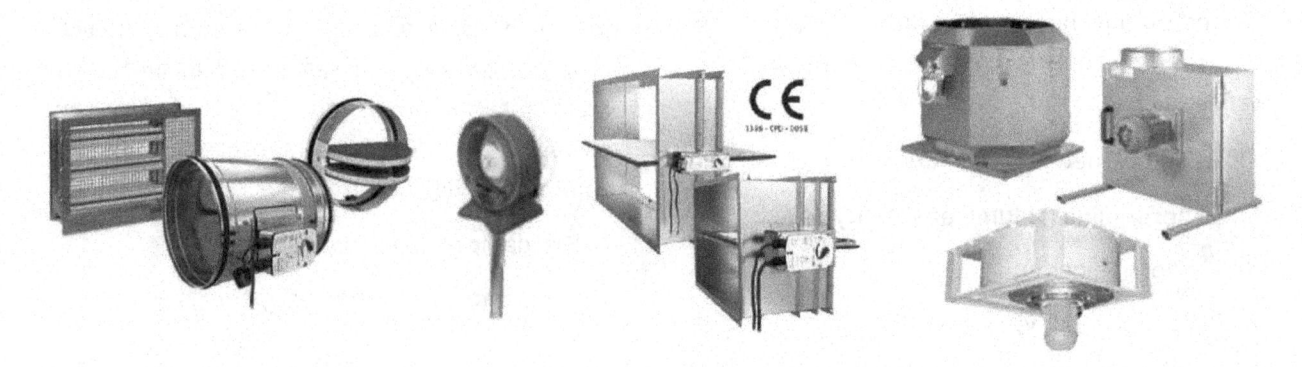

Figure 7.64. Typical components in fire and smoke management system from the left: fire dampers, smoke detector, smoke extract damper and smoke extract fan.

through a duct, transfer, or ventilation opening or 'active smoke control system' designed to control smoke migration using walls and floors as barriers to create pressure differences. Pressurizing the areas surrounding the fire prevents the spread of smoke into other areas. Smoke dampers have to be installed based on national building code. (Fig. 7.65.)

The commissioning involves verification that fire and smoke dampers as well as staircase pressurization and smoke extract fans actually operates as designed. Problems occur due to poor coordination between different contractors (civil, HVAC and BMS) as well as mutual ignorance of the installation requirements. These problems can be avoided by keeping proper records of installation inspections and commissioning checks, supported by the manufacturer's literature and fire simulation results.

Fire and smoke dampers <u>operation checks</u> include:

- Check for proper access to dampers for inspection & maintenance;

- Check that damper assemblies are installed as per the requirement & the manufacturer's instructions;

- Damper assembly is clean of any construction material;

- Check whether the fire damper is set to fully open position;

- Smoke damper position is either fully open or closed depending on the application;

- Check to ensure that the damper closure/ opening is not impeded;

- In case of motorized fire damper, verify operation in response to the activation of the fire alarm;

- In case of smoke dampers, verify operation in response to activation of smoke detectors;

- Ensure that dampers are tested according to sufficient standards.

The smoke extract fan needs to be pre-commissioned and tested as described in chapter 7.4 on exhaust fans. In addition, the response to activation of smoke detectors needs to be verified and that it can create high enough negative pressure to remove the smoke in case of fire.

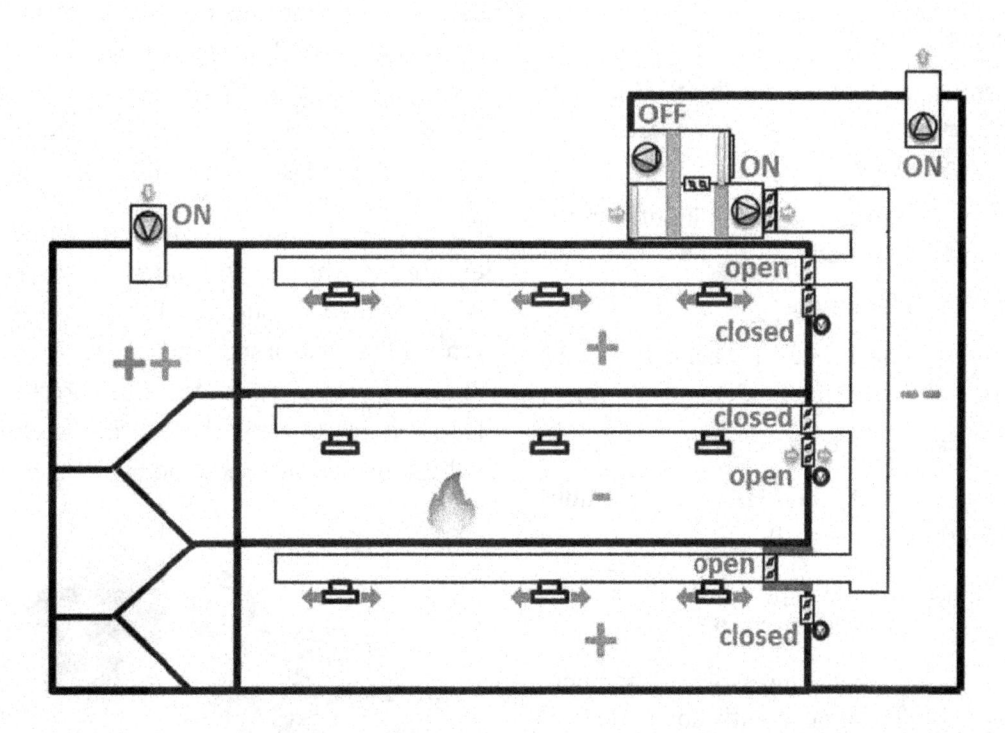

Figure 7.65. An overview of components in an active fire & smoke control system: fire dampers in supply ducts in the shaft wall (floor and shaft are different fire zones), motorized smoke dampers in each floor, smoke exhaust fan and staircase pressurization fan. Ground floor supply duct is fire insulated between the damper and wall. In case of fire, different parts of the system need to be open and closed. The main principle is to negatively pressurize the zone with fire and smoke extract paths and positively pressurize surrounding and evacuation n zones.

7.20. Staircase pressurization fan

Pressurization is required in staircases in buildings according to size and usage of the premises, to prevent smoke from exiting into the stairwell during the fire. Staircase pressurization fans (fig 7.66) are activated by a fire or smoke alarm in the building, but are shut down if smoke is detected as entering the fire exit.

These systems must develop a positive pressure (50Pa) within the exit to prevent smoke from entering while doors are closed, and when doors are open they must deliver an airflow velocity of 1m/s across every open doorway(s) to the fire-affected compartment.

Staircase doors must close and latch correctly. For smooth operational purposes, they must be manually opened inside to allow people to evacuate from the building. Therefore, there is a maximum pressure differential to be able to open all doors and minimum pressure differential to keep the smoke away from staircase. Staircase pressurization commissioning required comprehensive testing at every exit door and verification of the interface with a fire smoke detection system, which must activate and shut down the pressurization fans.

Pre-commissioning checks:

- Fan systems and dampers are installed as per the contract document and manufacturer's installation guidelines;

- Ensure that all fan and damper input/output points are wired correctly;

- Fire and smoke alarm systems are operational;

- All fan systems have been installed and pre-commissioned as per chapter 7.4;

- Electricity is available for operational checks

Start-up procedures of staircase pressurization system (initiate the event signal);

- Ensure all isolation dampers are open and supply fans switched ON;

- Measure and record pressure difference across each staircase door with doors closed; measured pressure difference should exceed value required by the National Building Code.

- Note that stack effect, wind speed and direction and outdoor temperature may all influence measured pressure and system balance. Therefore, record these parameters.

- Measure and record the force required to open one door using a spring scale.

- Hold the door open and measure the pressure differential across each staircase door again. Measure door opening force should not exceed the National Building Code, while the pressure differential across remaining doors should meet or exceed the code.

- Open the required number of doors one at a time, measuring and recording the force needed to open each door respectively, and the pressure differential across the remaining stairwell doors. Measure door opening force should not exceed the Building code, while the pressure differential across the remaining doors should meet or exceed code requirements.

- With required doors open, determine the direction of air flow across each door opening. Verify that air flows from stairwell to the occupied space.

Some pressurization systems use variable frequency drives (VFDs), sensors, relief dampers and apply a control sequence. The control sequence helps the system to adjust to any combination of door openings by maintaining the positive pressure differential across the opening. These systems compensate for changing conditions by either modulating air-flow or by releasing excess pressure from the staircase.

For such systems, carry out the above tests and ensure that the design staircase pressure set point is maintained throughout the test. The response time of the control sequence must also be checked. The response time of the pressurization control loop should not allow short term pressure value to fall below the value required by the code.

Figure 7.66. Staircase pressurization and smoke exhaust fan

Appendix A: COMMISSIONING CHECKLISTS

In the following pages there are commissioning templates of the most typical HVAC system components. For each product there is delivery checklist, pre-commissioning checklist, start-up procedures checklist as well as the operational data collection sheets. These are the checklists that contractor is requiring during the commissioning process. There are two separate checklists for the client and commissioning provider to perform the installation and operation reviews.

Table 8. The following 67 checklists are presented in this appendix.

Checklists for HVAC Contractor

AHU/TFA Delivery Check List				
Project:			**AHU/TFA name:**	
Contractor:			**AHU/TFA ID #:**	
SI. No			Value	Remarks, if not as per specification
1	Manufacture			
2	Unit dimesions			
3	Site assembled or packaged			
4	System type (constant air volume, variable air volume (VAV), energy recovery etc.)			
5	Cooling coil type (chilled water or DX)			
6	Cooling coil dimensions			
7	Slope towards drain in the stainless steel drain pan			
8	Rated cooling capacity			
9	Heating coil type (hot water or electrical)			
10	Heating coil dimensions			
11	Rated heating capacity			
12	Orientation of cooling and heating coil connections			
13	Supply Fan model and type (EC, AC, backward curved, etc)			
14	Air flow rate			
15	Fan motor power			
16	Number of fans per unit			
17	Fan belt type, size and number			
18	Vibration eliminators			
19	Orientation of fan section doors			
20	Energy recovery wheel (size, depth, motor, material, etc.}			
21	Filter types			
22	Filter sizes and quantity			
23	Orientation of filter section doors			
24	All factory-build measurement instruments in place;			
25	All factory·build electrical connections done;			
26	Volume control damper for supply/return/fresh air duct connections			
27	Canvas connection in the fan outlet			
28	Was the 'Authorized Person' in charge to receive goods trained to receive the delivery and is aware of the layout of customers facilities and the customers off load procedure?			
29	Test & warranty certificates for motor/coils/wheels/fan			
30	Unit checked for interior and exterior damages			
31	AHU/TFA received as ordered			

Checked by:

Date:

	AHU/TFA Pre Commissioning Check list			
Project:			**AHU/TFA name:**	
Contractor:			**AHU/TFA ID #:**	
Sl. No			Yes/No/NA	Remarks
1	Delivery checks of AHU/TFA are completed			
2	Correct AHU/TFA is installed and located as per drawing			
3	Clear access to all required components			
4	All transit blocks are removed			
5	Supply & return ducts are connected and complete			
6	Sufficient rain hood and bird net is installed in case unit is located outside.			
7	Canvas between unit a nd ductwork is installed correctly			
8	Ensure that air flow and static pressure measurement sensors are located in a straight section of duct (not right after damper, branch or damper)			
9	All volume control dampers are in fully open position (in case ductwork is not yet balanced)			
10	All pipe work connections are completed to all coils			
11	All heating & cooling coils are secured and fins combed			
12	Pipe supports are sufficient			
13	All valves, gauges and measurmeent instruments are installed			
14	Proper condensate drain piping is installed with slopes and supports			
15	Chilled and hot water pipe insulation work is done as per design drawings			
16	All filters are as per design and installed correctly			
17	Temporary filter media is installed covering air intake and return connections to protect main filters			
18	Fans and motors rotates properly to the right direction			
19	Fan belts are fitted correctly			
20	Fan shaft and bearings are aligned			
21	Cleanliness of the bearings, lubricant is fresh and of the correct grade			
22	Coolant is a vailable at the bearings (if specified)			
23	Drive guards are fitted and access for speed measurement is provided			
24	Pulleys are aligned and belt tension is correct			
25	Any anti vibration mountings required are installed			
26	All electrica l isolators are in place and connections complete and tested to motors			
27	Electrical cabling connections are with cable tray			
28	All sensors and controls are installed, calibrated and fully operational			
29	Variable Frequency Driver (VFD) is installed and operational			
30	Blanking of gaps a re done where necessary with sealant			
31	Lighting within units is in place and connected			
32	Dampers are free moving through full range			
33	Appropriate safety warning signs are in place			
34	AHU is free from damages/scratches			
35	All components, ducts and pipes as well as AHU rooms are clean			
36	Power is available during commissioning			
37	BMS data is available during the commissioning			
38	Pressure testing of unit has been satisfa ctorily completed. Factory assembled units are tested on factory and site assembled on site			
39	AHU/TFA system is in operation before the test and balance procedures are started			
40	AHU/TFA manufacturer has pressure tested the unit after assambly on site or in case of factory assambled packaged unit, the pressure testing certificate has been handed over to the contractor			
Checked by:				
Date:				

AHU/TFA Start Up Check List			
Project:		**AHU/TFA name:**	
Contractor:		**AHU/TFA ID #:**	
Sl. No		Yes/No/NA	Remarks
1	AHU/TFA pre-commissioning checks are completed		
2	Electrical tests are complete & available from electrical contractor		
3	Ensure that the ductwork is properly balanced before making operational measurements in AHU/TFA and provide air flow measurement tables and ratio calculations		
4	Ensure that static pressure control of fan operates properly		
5	Ensure that all volume control damper positions are recorded and marked to the dampers		
6	Check the fan for proper operating condition and ensure that the fan motor is below the full-load current		
7	Measure the air flow rates in all main ducts and confirm that the total discrepancy is within 10%		
8	Determine fan air flow by measuring the air velocity in front of the filter section. Air flow rate should be 100? 110% of designed air flow rate - compare the air flow rate also with the results on the main ducts after the AHU to ensure there is no air leakage		
9	Measure static pressure difference across the fan section (pressure creation) and across all other components in the AHU/TFA (pressure loss)		
10	Define the operation point of fan in the fan curve and attach it into the commissioning documents		
11	Check the fan and motor frequency, voltage, currency and RPM		
12	Measure the cooling and heating water systems operation parameters		
13	Measure the operation of energy recovery wheel, air temperatures and flow rates both in supply and exhaust side and calculate the energy recovery efficiency		
14	Measure the rotation speed of a wheel and compare it to the selection details		
15	Measure outdoor, supply and return air quality parameters		
16	Check the air diffusion of all diffusers so that they do not create local draught		
17	AHU/TFA control and operation algorithms are correctly configured in BMS		
18	Simulate AHU/TFA operations (start/stop) and review the operation of related components (e.g. motorized dampers, cooling coil valves, exhaust fans, smoke and fire dampers, staircase pressurization fan etc.)		
19	AHU/TFA is operating as specified in Sequence of Opreration (SOO)		

Checked by:

Date:

	AHU/TFA Operation Data Collection Sheet				
Project:			**AHU/TFA name:**		
Contractor:			**AHU/TFA ID #:**		
Sl. No		Unit	Designed	Measured	Difference
1	Supply air flow rate (balancing data collection sheets in ductwork section)				
2	Pressure generation of fan with maximum air flow rate				
3	Static pressure across all filters				
4	Static pressure across cooling coil				
5	Static pressure across Energy Recovery Wheel				
6	Static pressure across volume control dampers				
7	Supply air flow rate with all VAV dampers in min and max positions+AI:F22 (each VAV damper measurement in separate sheet)				
8	VFD/fan motor frequency with maximum air flow rate				
9	VFD frequency with minimum air flow rate				
10	Fan RPM with maximum air flow rate				
11	Fan motor RPM with maximum air flow rate				
12	Fan motor voltage with maximum air flow rate				
13	Fan motor current with maximum air flow rate				
14	Center-to-center distance of the motor base travel				
15	Supply air volume control damper (VCD) position				
16	Fresh air flow rate				
17	Fresh air VCD damper position				
18	Return air flow rate				
19	Return air VCD damper position				
20	Exhaust air flow rate (DOAS unit)				
21	Outdoor air temperature				
22	Supply air temperature (cooling coil valve fully open)				
23	Outdoor air relative humidity				
24	Supply air relative humidity (cooling coil valve fully open)				
25	Supply air temperature after Energy Recovery Wheel				
26	Exhaust air temperature after Energy Recovery Wheel				
27	Energy Recovery Wheel rotating speed				
28	Inlet water temperature of cooling coil				
29	Return water temperature of cooling coil				
30	Water flow rate of cooling coil				
31	Inlet water temperature of heating coil				
32	Return water temperature of heating coil				
33	Water flow rate of heating coil				
34	Heating capacity of electrical heater				
35	Noise level of AHU/TFA				
36	Outdoor air PM 10 concentration				
37	PM10 concentration after all filters and fan section				
38	Outdoor air PM2.5 concentration				
39	PM2.5 concentration after all filters and fan section				
40	Out door air NO_2 concentration				
41	NO_2 concentration after all filters and fan section				
42	Outdoor air O_3 concentration				
43	O_3 concentration after all filters and fan section				
44	Out door air CO concentration				
45	CO concentration after all filters and fan section				

Checked by:

Date:

AHU/TFA Airflow Measurement - Before & After Balancing											
Project:			**AHU/TFA name:**								
Contractor:			**AHU/TFA ID #:**								
Number of filter	Filter size (mm) (e.g. 600x600}	velocity 1	velocity 2	velocity 3	velocity 4	velocity 5	Average	Filter Area	Air flow rate		
				m/s			m/s	m2	m3/s	CFM	
1											
2											
3											
4											
5											
6											
7											
8											
9											
10											
								Total:			
							Designed air flow rate:				
							Difference:				

Fan speed with maximum air flow rate: [   ] Hz

Checked by:
Date:

Exhaust Fan Delivery Check List			
Project:		**Fan name:**	
Contractor:		**Model ID#:**	
Sl. No	Description	Value	Remarks, if not meeting the specificat ion
1	Manufacturer		
2	Model		
3	Fan type		
4	Air flow rate		
5	Static pressure		
6	Fan model and type (EC, AC, backward curved, et c)		
7	Fan motor power		
8	Fan belt type, size and number		
9	Vibration eliminators		
10	Test & Warranty certificates for motor/fan		
11	Unit checked for interior and exterior damages		
12	Was the "Authorized person" in charge of receiving goods trained to receive the delivery and is aware of the layout of customers facili ties and the customers off load procedure?		
13	Fan received as ordered		

Checked by:
Date:

\multicolumn{5}{c}{**Exhaust Fan Pre-commissioning Check List**}
Project:			**Fan name:**	
Contractor:			**Model ID#:**	
Sl. No	Description		Yes/No/NA	Remarks
1	Delivery checks of all fans were made			
2	Correct fan is installed and located as per drawing			
3	Safety disconnect installed in an accesible location			
4	All transit blocks are removed			
5	Clear access to fan			
6	All electrical connections are tight and correctly wired			
7	Fan rotation is in proper direction			
8	Duct is properly connected with the canvas			
9	Fans and motors rotates properly to the right direction			
10	Fan belts are fitted correctly			
11	Fan shaft and bearings are aligned			
12	Pulleys are aligned and belt tension is correct			
13	Cleanliness of the bearings, lubricant is fresh and of the correct grade			
14	Coolant is available at the bearings (if specified)			
15	Drive guards are fitted and access for speed measurement is provided			
16	Ensure satisfactory operation of the inlet guide vanes over full range of movement			
17	Check that fan casings is earthed correctly and soundly bonded			
18	Any anti vibration mountings required are installed			
19	All electrical isolators are in place and connections complete and tested to motors			
20	Power is available during commissioning			
21	BMS data is available during commissioning			
22	Fan systems is in operation before the test and balance procedures are started			

Checked by:

Date:

\multicolumn{5}{c}{**Exhaust Fan Start-up Check List**}
Project:			**Fan name:**	
Contractor:			**Model ID#:**	
Sl. No	Description		Yes/No/NA	Remarks
1	Exhaust fan pre-commissioning checks are completed			
2	Electrical tests are complete & available from electrical contractor			
3	Check the fan for proper operating condition and ensure that the fan motor is below the full-load current			
4	Measure the air flow rates in main ducts and confirm that the total discrepancy is within 10%			
5	Measure static pressure difference across the fan section (pressure creation)			
6	Define the operation point of fan in the fan curve and attach it into the commissioning documents			
7	Check the fan and motor frequency, voltage, currency and RPM			

Checked by:

Date:

Exhaust Fan Operation Data Collection Sheet

Project:			Fan name:		
Contractor:			Model ID#:		
SI. No		Unit	Designed	Measured	Difference
1	Supply air flow rate (main duct and diffuser values in separate sheet)				
2	Pressure generation of fan with maximum air flow rate				
3	Fan RPM with maximum air flow rate				
4	Fan motor RPM with maximum air flow rate				
5	Fan motor voltage with maximum air flow rate				
6	Fan motor current with maximum air flow rate				
7	Noise level of fan				
8	Center·to·center distance of the mot or base travel				

Checked by:

Date:

VFD Delivery Check List

Project:			Fan name:	
Contractor:			Model ID#:	
SI. No		Value	Remarks, if not meeting the specificat ion	
1	Manufacturer			
2	Model no.			
3	Customer connection digram is attached			
4	Project specific schematic diagram attached			
10	Test & warranty certificates			
11	Unit checked for interior and exterior damages			
12	Was the 'Authorized Person' in charge of receiving goods trained to receive the delivery and is aware of the layout of customers facilities and the customers off load procedure?			
13	VFD received as ordered			

Checked by:

Date:

VFD Pre-commissioning Check List

Project:			Fan name:	
Contractor:			Model ID#:	
SI. No		Yes/No/NA	Remarks	
1	Delivery checks of all VFDs were made			
2	Correct VFO is installed with each fan			
3	Ensure that drive, power line and motor are at the same voltage			
4	Ensure that rated drive current is greater than or equal to the total full load current of all motors which will be driven at once.			
5	A disconnect or contactor between the drive and the motor need to be interlocked to the drive			
6	Ensure that multiple motors have appropriate and individual motor overload and short circuit protection			
7	There are no power factor correction capacitors between the drive and the motor.			
8	Two speed motors are wired permanently for full speed.			
9	Y·start, ?-run motors are wired permanently for run.			
10	Part winding start motors are wired permanently for run.			
11	Check that all wiring connections are secure.			
12	Ensure that each drive is grounded individually, no daisy chain grounds.			
13	0·10 V DC and mA signal wires protected from noise.			
14	There are separated runs for input power, motor power and control wiring.			
15	Power is available during commissioning			
16	BMS data is available during commissioning			
17	VFO is in operation before the testing and balance procedures are started			

Checked by:

Date:

VFD Start-up Check List			
Project:		**Fan name:**	
Contractor:		**Model ID#:**	
Sl. No	Description	Yes/No/NA	Remarks
1	VFD pre-commissioning checks are completed		
2	Measure phase to phase line voltage and ensure measured voltage is within drive specification		
3	Measure phase to ground voltage. If any measured phase voltage is greater than 60% of phase to phase Voltage, open RFI switch		
4	Double check all wire connections (correct terminal connection, correct tightness)		
5	Power the drive: ensure that all run commands are off, all speed commands are set to zero - switch power on - display and power LED gets on		
6	Set all the parameters as described in the VFD manual		
7	Manual mode: check the motor's rotation from the drive. If incorrect, reverse two leads between the drive and the motor.		
8	Manual mode: if a bypass is provided, check the motor's rotation in bypass mode. If incorrect, reverse two input power lines.		
9	Manual mode: accelerate the motor to full speed and verify operation.		
10	Manual mode: decelerate the motor to a stop and verify operation.		
11	Manual mode: slowly operate the motor over the speed range and check for resonance.		
12	There is no unusual noise.		
13	Automatic mode: ensure that the drive follows run/stop commands from the system.		
14	Automat ic mode; ensure that the drive follows the speed command from the system.		
15	Automatic mode; set final min/max references as required		
16	Automatic mode: set final ramp up/down times as required		
17	Automatic mode: set up the PIO control parameters as required.		

Checked by:

Date:

VFD Data Collection Sheet				
Project:		**Fan name:**		
Contractor:		**Model ID#:**		
Sl. No	System Parameter	Unit	Designed	Measured	Difference
1	Line voltage - phase 1				
2	Line voltage - phase 2				
3	Line voltage - phase 3				
4	Phase to ground voltage				
5	Final min reference				
6	Final max reference				
7	Final ramp-up time				
8	Final ramp down time				

Checked by:

Date:

	Ductwork Pre-commissioning Check List		
Project:		Contractor	
Sl. No	Description	Yes/No/NA	Remarks
1	Ductwork is completed		
2	Was the "Authorized person" in charge of receiving goods who is trained to receive the delivery and is aware of the layout of customers facilities and the customers off load procedure?		
3	Required thickness of sheet has been used in ducting as per duct size		
4	Duct sheets (GI/pre-insulated) have been checked when arriving to site and stored sufficiently to ensure high quality of material		
5	There is a uniform cross-section area through out the duct length		
6	All bents, T-branches, reductions and expansions are made so that they do not increase unnecessarily pressure loss or create the additional leakage		
7	The hole in the main duct is the same size than e.g. the T-branch duct connection		
8	There are no sharp edges inside the ductwork		
9	All joints are sequered with aluminium duct tape		
10	All flanges are properly made and gasget has been used in joints to ensure air tight installation		
11	Duct is sufficiently supported (type, distance, et c.)		
12	There is no dents or other damages in ducts		
13	Confirm duct work runs are correctly located and have adequate clearance from other services		
14	Ductwork leakage test (as specified in design documents) was completed before insulation		
15	All Volume Control Dampers (VCD), VAV dampers and static pressure dampers are correctly installed (location, safety distance, direction, access to actuator)		
16	Fire dampers, smoke dampers, and access doors are installed as designed		
17	Fire insulation around duct is sufficient and fire/smoke tight as per fire rating, where specified		
18	All dampers, motors and sensors are reachable		
19	Sound attenuators are properly made and do not release Man-made Vitorous Fibres (MMVF) into the air		
20	Technical data of insuration material is available and turms the specifiaction in terms of internal conductivity, water vapour resistance, surface emissivity and insulation protection from ambient conditions like solar, temperatire and rain		
21	Insulation is properly made (fully air tight, no damages in suriace, no air pockets between insulation and		
22	Insulation material is protected from weather (sun & rain) in outdoor installations		
23	As-built drawings has been submitted		
Checked by:			
Date:			

Ductwork start-up Check List			
Project:		**Contractor:**	
SI. No		Yes/No/NA	Remarks
1	Ensure that all pre-commissioning checks has been made		
2	Leakage testing of ductwork was carried out before insulation		
3	The ductwork leakage test reports are available		
4	Ensure that all ducts are dean inside, if not, carry out the ductwork cleaning		
5	Ensure that all VCD and VAV dampers are in fully open position and TFA/AHU air flow rate is 110-120% of designed air flow rate before starting the ductwork balancing		
6	Adjust air diffusion in the space if it generates too high velocity in the occupied zone		
7	Measure all air flow rates and calculate the ratios between measured and designed air flow rates In each diffuser/grille and VCD damper		
8	Carry out the proportional balancing of each diffuser/grille damper, sub- and main branch VCD dampers and in the end reduce the fan speed to achieve the designed air flow rate		
9	Measure all air flow rates (each diffuser/grille, VCD dampers) and record them as a final balancing document together with the records of each damper position		
10	Ensure that there is not too much noise generated by the ventilation system		
11	Measure the air flow rate of each VAV damper with maximum and minimum position after balancing work is done		

Checked by:

Date:

Ductwork Balancing/Ratio Calculations and Final Measurement						
Project:		**AHU/TFA name:**				
Contractor:		**AHU/TFA ID #:**				
Zone	Item ID	Before Balancing			After Balancing	
		Air flow rate (dm3/s)		Measured/ Designed Ratio	Air flow rate (dm3/s)		Measured/ Designed Ratio
		Designed	Measured		Designed	Measured	

Checked by:

Date:

Ductwork Balancing/Ratio Calculations and Final Measurement								
Project:				**Contractor:**				
Ductwork ID#	Duct surface area	Static pressure	Differential pressure	Leakage factor	Permitted leakage	Actual duct leak	Total area CFM	% duct leakage (<3%)
	ft2	Pa	Pa	CFM/100112	CFM	CFM	CFM	

Checked by:

Date:

	Staircase Pressurization Fan Delivery Check List		
Project:		Fan name:	
Contractor:		Model ID#;	
SI. No		Value	Remarks, if not as specified
1	Manufacturer		
2	Model		
3	Fan type		
4	Airflow rate		
5	Static pressure		
6	Fan model and type (EC, AC, backward curve, etc)		
7	Fan motor power		
8	Fan belt type, size and number		
9	Vibration eliminators		
10	Test & Warranty certificates for motor/fan		
11	Unit checked for interior and exterior damages		
12	Was the "Authorized person" in charge of receiving goods who is trained to receive the delivery and is aware of the layout of customers facilities and the customers off load procedure?		
13	Fan received as ordered		

Checked by:

Date:

	Staircase Pressurization Fan Pre-commissioning Check List		
Project:		Fan name:	
Contractor:		Model ID#;	
SI. No	Description	Yes/No/NA	Remarks
1	Delivery checks of all fans were made		
2	Fan systems & dampers are installed as per the contract document & manufacturer's installation guidelines		
3	Correct fan is installed and located as per drawing		
4	Safety disconnect installed in an accesible location		
5	All transit blocks are removed		
6	Clear access to fan		
7	All electrical connections are tight and correctly wired		
8	Fan rotation is in proper direction		
9	Duct is properly connected with the canvas		
10	Fans and motors rotates properly to the right direction		
11	Fan belts are fitted correctly		
12	Fan shaft and bearings are aligned		
13	Pulleys are aligned and belt tension is correct		
14	Cleanliness of the bearings, lubricant is fresh and of the correct grade		
15	Coolant is available at the bearings (if specified)		
16	Drive guards are fitted and access for speed measurement is provided		
17	Ensure satisfactory operation of the inlet guide vanes over full range of movement		
18	Check that fan casings is earthed correctly and soundly bonded		
19	Any anti vibration mountings required are installed		
20	All electrical isolators are in place and connections complete and tested to motors		
21	Power is available during commissioning		
22	Fire and smoke alarm systems are operational		
23	Fan systems is in operation before the test and balance procedures are started		

Checked by:

Date:

Staircase Pressurization Fan Start-up Check List

Project:		Fan name:	
Contractor:		Model ID#;	
Sl. No	Description	Yes/No/NA	Remarks
1	Exhaust fan pre-commissioning checks are completed		
2	Electrical tests are complete & available from electrical contractor		
3	Ensure all isolation dampers are open and supply fans switched ON.		
4	Check the fan for proper operating condition and ensure that the fan motor is below the full-load current		
5	Measure the air flow rate and confirm that the total discrepancy is within 10%		
6	Measure static pressure difference across the fan section (pressure creation)		
7	Check the fan and motor frequency, voltage, currency and RPM		
8	Measure and record pressure difference across each staircase door with doors closed. Measured pressure difference should exceed value required by the national building code.		
9	Record wind speed & direction and outdoor temperature		
10	Hold the door open and measure the pressure differential across each staircase door again. Measured door opening force should not exceed the national building code, while the pressure differential across remaining doors should meet or exceed the code (13.5 Pa)		
11	If specified in CX plan, open the required number of doors one at a time, measuring and recording the force needed to open each door respectively, and the pressure differential across the remaining stairwell doors. Measured door opening force should not exceed the building code, while the pressure differential across the remaining doors should meet or exceed code requirements.		
12	With required doors open, determine the direction of airflow across each door opening. Verify that air flows from stairwell to the occupied space.		

Checked by:

Date:

Staircase Pressurization Fan Operation Data Collection Sheet

Project:			Fan name:				
Contractor:			Model ID#;				
Sl. No	System Parameter	Unit	Designed	Measured	Difference		
1	Supply air flow rate (main duct and diffuser values in separate sheet)						
2	Pressure generation of fan with maximum air flow rate						
3	Fan RPM with maximum air flow rate						
4	Fan motor RPM with maximum air flow rate						
5	Fan motor voltage with maximum airflow rate						
6	Fan motor current with maximum airflow rate						
7	Center-to-center distance of the motor base travel						
8	Outdoor air temperature						
			Door 1	Door 2	Door 3	Door 4	Door 5
9	Pressure difference across each door/all doors closed						
10	Required force to open the door						
11	Pressure difference across each door/door 1 open, doors 2,3,4,5 closed						
12	Required force to open the door						

Checked by:

Date:

	Chiller (Vapour Compression) Delivery Check List		
Project:		**Chiller name:**	
Contractor:		**Chiller ID#:**	
Sl. No	Description	Value	Remarks, if specification is not met
1	Manufacturer		
2	Chiller model no.		
3	Evaporator model no. if not included in chiller no.		
4	Condender model no. if not included in chiller no.		
5	Chiller compressor type (centrifugal, screw, reciprocating, scroll)		
6	Condenser type (water or air cooled)		
7	Compressor motor voltage		
8	Locked Rotor Amps {LRA}: current under starting conditions		
9	Rated Loads Amps (RLA): maximum current under any operating condition		
10	All accessories as specified		
11	Pipe connections are correct in size		
12	Refrigerant type		
13	Manufacturer's efficiency rating (kW/ton) in standard conditions and IEER		
14	Was the 'Authorized Person' in charge to receive goods trained to receive the delivery and is aware of the layout of customers facilities and the customers off load procedure?		
15	Test & warranty certificates		
16	Unit checked for exterior damages		
17	Chiller received as ordered		

Checked by:

Date:

	Chiller (Vapour Compression) Pre-commissioning Check List		
Project:		**Chiller name:**	
Contractor:		**Chiller ID#:**	
Sl. No		Yes/No/NA	Remarks
1	Delivery checks of chiller are completed		
2	Correct chiller is installed and located as per drawing		
3	All pipe connections are completed		
4	Clear access to all required components		
5	Factory start-up sheet completed and attached		
6	Test and balance report is reviewed for chiller system flows		
7	All accessories (e.g. pressure gauges and thermometers) are on place		
8	That expansion valve bulbs, and any other temperature or pressure sensing bulbs are correctly located with capillary tubes free from damage or distortion		
9	That pipeline tapings (with fixed or test gauges) for pressure and temperature measurements are available on chilled water and condenser water circuits		
10	Chiller and surrounding areas are clean		
11	No visible water or oil leaks		
12	No unusual noise or vibration		
13	Vibration insulators are on place		
14	Appropriate safety warning signs in place		
15	Refrigerant fill is sufficient and refrigerant is clean from dirt and moisture		
16	Confirm power availability for duration of commissioning		
17	BMS data is available during the commissioning		
18	Chiller system is in operation before the test procedures are started		

Checked by:

Date:

	Chiller {Vapour Compression) Start-up Check List		
Project:		**Chiller name:**	
Contractor:		**Chiller ID#:**	
Sl. No		Yes/No/NA	Remarks
1	Chiller pre-commissioning checs are completed		
2	Electrical test sheets are completed and available from electrical contractor		
3	All chiller accessories (high and low pressure limits, freezing limit, Hquid receiver, filter drier, refrigerant sight glass, liquid solenoid valve, hot gas by-pass valve, water makeup solenoid valve, evaporator flow switch, etc.) are performing properly		
4	Measure all chiller operation parameters {temperatures, pressures, flow rates, electrical parameters}		
5	Ensure that there is no water or bubbles in the refrigerant		
6	Chiller operates without unusual number of trips		
7	Outside air temperature lockout functions properly		
8	Controlling independent variable is set correctly for chilled water temperature reset		
9	Chilled water temperature follows reset schedule		
10	Ensure that chiller maintains chilled water temperature set point ± 1 oC over a 2 hour operating period		
11	Remove the chiller load and verify the chiller and accessory shutdown sequence		
12	Add load and verify the chiller and accessory start -up sequence		
13	Shutdown and start-up sequences stage multiple chillers & accessories properly		
14	Carry out Running In procedures and related inspections afterwards.		
15	Chiller control algorithm operates as specified in Sequence of Operations (SOO)		
16	Chiller control and operation algorithms are correctly configured in BMS		
17	Chiller appears to meet load		

Checked by:

Date:

	Chiller Operation Data Collection Sheet						
Project:			**Chiller name:**				
Contractor:			**Chiller ID#:**				
Sl. No	System Parameter	Unit	Designed	Measured	Difference	Make	Break
1	Out door air temperature						
2	Evaporator water temperature in						
3	Evaporator water temperature out						
4	Condenser water temperature in						
5	Chilled water temperature during 2 h period min						
6	Chilled water temperature during 2 h period max						
7	Condenser water temperature out						
8	Evaporator water flow rate						
9	Condenser water flow rate						
10	Air cooled compressor: air temperature in						
11	Air coaled compressor: air temperature out						
12	High pressure						
13	Low pressure						
14	Number of trips per hour						
15	Chiller control valve/Balancing Vavle position						
16	Condenser control valve/Balacing Valve position						
17	locked Rotor Amps (LRA): current under starting conditions						
18	Rated Loads Amps (RLA): maximum current under any operating condition						
19	Voltage - phase I						
20	Current - phase I						
21	Speed - phase I						
22	Voltage - phase II						
23	Current - phase II						
24	Speed - phase II						
25	Voltage - phase III						
26	Current - phase III						
27	speed - phase III						
28	Voltage imbalance (<2%)						
29	Chiller noise level (at lm apart)						

Checked by:

Date:

Pump Delivery Check List

SI. No		Value	Remarks in not as per specification
Project:		**Pump name:**	
Contractor:		**Pump ID#:**	
1	Manufacture		
2	Type of the pump		
3	Application (chilled water, cooling tower, potable water etc.)		
4	Water flow rate		
5	Head		
6	Pump motor		
7	Motor Voltage		
8	Motor Power		
9	Motor rpm		
10	Rated current of motor		
11	Rated motor efficiency		
12	Minimum circulation required		
13	Type of flow control system installed		
14	Vibration eliminator		
15	Test & Warranty certificates		
16	Pipe connections are correct in size		
17	Pump checked for exterior damages		
18	Was the 'Authorized Person' in charge of receiving goods trained to receive the delivery and is aware of the layout of customers facilities and the customers off load procedure?		
19	Pump received as o rdered		

Checked by:

Date:

Pump Pre-commissioning Check List

SI. No		Value	Remarks
Project:		**Pump name:**	
Contractor:		**Pump ID#:**	
1	Delivery checks of all pumps were made		
2	Correct pump is installed and located as per drawing		
3	There is a clear access to pump		
4	All electrical connections are tight and correctly wired		
5	Motors and pumps are aligned properly		
6	Impeller is free to rotate when decoupled and should not make unusual noise when rotated by hand		
7	Bearings are clean and lubricated and drive guards are fitted		
8	Pressure test points are provided at the suction and discharge of pump		
9	Flow measurement devices installation as per approved shop drawing		
10	System has been flushed and cleaned		
11	Pump strainer is cleaned after flushing		
12	Pump drain is connected to main drain		
13	Triple duty valve (if specified) is provided at pump discharge		
14	Flexible connectors are installed as specified		
15	Check that the direction sign of an non-return valves is among the same discharge direction of associated pumps		
16	Check that the horizontal or vertical alignment of all flexible joints is within the tolerance recommended by manufact urer's installation guideline		
17	Pump is insulated (if specified)		
18	Pump and motor are lubricated		
19	Suction diffuser (if specified) is installed at pump suction		
20	VFD is installed and all controls connected		
21	Installation of inertia base and vibration isolation springs		
22	Ensure system is free of foreign matter which could damage the pump.		
23	Power is available during commissioning		
24	BMS data is available during commissioning		
25	Pump systems is in operation before the test and balance procedures are started		

Checked by:

Date:

Pump Startup Check list

Project:			Pump name:	
Contractor:			Pump ID#:	
Sl. No			Yes/No/NA	Remarks
1	Pump pre-commissioning checks are completed			
2	Electrical tests are complete & available from electrical contractor			
3	Check the pump for proper operating condition and ensure that the pump motor is below the full-load			
4	Pump rotation is verified			
5	Factory alignment is checked			
6	Check that all normally open isolating and regulating valves are fully open and that all normally closed valves are dosed			
7	Manually turn coupling to assure free rotation of pump and motor.			
8	Open all control valves to full flow to heat exchangers of branch circuits			
9	Fully open the return and close the flow valve on the pump.			
10	Close valves on standby pump. Closing the flow valve on the duty pump will limit the initial starting current, which is usually excessive at the first time a pump is running due to bearings stiffness.			
11	Measure the water flow rates in the pump and confirm that the discrepancy is within 10%			
12	Measure the pump head (pressure creation)			
13	Define the operation point of pump in the pump curve and attach it into the commissioning documents			
14	Ensure that there is no unusual noise of pump			
15	Check the pump and motor frequency, voltage, currency and RPM			
16	Pump operates as per SOO			

Checked by:

Date:

Pump Operation Data Collection Sheet

Project:			Pump name:		
Contractor:			Model ID#:		
Sl. No		Unit	Designed	Measured	Difference
1	Water flow rate				
2	Pump suction head				
3	Pump discharge head				
4	Pump speed (rpm) with maximum water flow rate				
5	Pump motor voltage with maximum water flow rate				
6	Pump motor current with maximum water flow rate				
7	Pump noise level				
8	Supply voltage				
9	Starting current				
10	Running current				
11	Overload Setting				
12	Tripping Time of St arter Overload				

Checked by:

Date:

Cooling Tower Delivery Check List			
Project:		**Name:**	
Contractor:		**ID#:**	
SI. No		Value	Remarks in case not as per specification
1	Manufacturer		
2	Model no.		
3	Coolina tower tvoe (open circuit/closed circuit)		
4	Number fans		
5	Fan type		
6	Drift eliminator on place		
7	Fill is as specified		
8	Pipe connections are correct in size		
9	Was the 'Authorized Person' in charge to receive goods trained to receive the delivery and is aware of the layout of customers facilities and the customers off load procedure?		
10	Test & warranty certificates		
11	Unit checked fo r exterior damages		
12	Cooling tower received as ordered		

Checked by:

Date:

Cooling Tower Pre-commissioning Check List			
Project:		**Name:**	
Contractor:		**ID#:**	
SI. No		Yes/No/NA	Remarks
1	Delivery checks of cooling tower are completed		
2	Correct cooling tower is installed and located as per drawing		
3	All pipe connections are completed		
4	Clear access to all required components		
5	The water-circulating system serving the cooling tower is cleaned		
6	That interior filling of cooling tower is clean and free of foreign materials such as scale, algae and tar		
7	Fill is fixed properly and clean		
8	Drift is fixed properly and clean		
9	The cooling tower fans are free to rotate and the tower basin is clean		
10	There is no unnormal noise or vibration		
11	Vibration insulators are on place		
12	Fans are fixed properly		
13	Drive alignment and belt tension has been checked		
14	Bearings & lubrication has been checked		
15	Drainage & fall		
16	Strainer is clean		
17	Ball valve function		
18	Water level in the tower basin is maintained at the proper level,		
19	Water treatment equipment		
20	Water make-up solenoid		
21	Electrical supply connections		
22	Earth bonding		
23	Confirm power availability for duration of commissioning		
24	BMS data is available during the commissioning		
25	Cooling tower is in operation before the test procedures are started		

Checked by:

Date:

	Cooling Tower Start-up Check List			
Project:			**Name:**	
Contractor:			**ID#:**	
SI. No			Yes/No/NA	Remarks
1	Cooling tower pre-commissioning checks are completed			
2	Electrical test sheets are completed and available from electrical contractor			
3	Verify that the minimum speed settings on VSDs have been coordinated with the requirements of the cooling tower drive train			
4	Ensure, that centrifugal action during full flow does not cause any entrainment of air which may cause pump cavitation			
5	All valves except balancing valves in the water system are in full open position			
6	Verify the sensor that the control system used to sequence the fans & bypass valve is located downstream of the point where the bypass water mixes with the water coming from the tower basins			
7	Verify all safety interlocks work properly.			
8	Verify system & control loop performance & stability at all load conditions (cool & dry, hot & humid) and flow conditions (min and max air and water flow rate)			
9	Verify that basin heaters & heaters systems are shut down when the temperatures rise above freezing			
10	Verify that seismic restraints have been properly installed & adjusted			
11	Verify that start-up screens have been removed from the strainers & replaced with gravel screens			
12	Shutdown and start-up sequences stage multiple cooling towers perform properly			
13	Use entering and leaving wet bulb temperatures to determine the tower performance			
14	Use entering and leaving wet and dry bulb temperatures to determine the rate of evaporation			
15	Cooling tower water sample has been taken and analysed			
16	Cooling tower water quality is acceptable			
17	Cooling tower control algorithm operates as specified in Sequence of Operations (S00)			
18	Cooling tower operates as per design			

Checked by:

Date:

	Cooling Tower Operation Data Collection Sheet				
Project:			**Name:**		
Contractor:			**ID#:**		
SI. No	System Parameter	Unit	Designed	Measured	Difference
1	Ambient air dry bulb temperature				
2	Ambient air wet bulb temperature				
3	Entering air dry bulb temperature				
4	Entering air wet bulb temperature				
5	Leaving air dry bulb temperature				
6	Leaving air wet bulb temperature				
7	Cooling water flow rate				
8	Cooling water entering temperature				
9	Cooling water leaving temperature				
10	Make up water flow rate				
11	Constant bleed water flow rate				
12	Fan air flow rate				
13	Fan power				
14	Fan pressure				
15	Supply voltage				
16	Motor starting current				
17	Motor running current				
18	Motor speed				
19	Fan speed				

Checked by:

Date:

Pipework Start-up Check List			
Project:		**Contractor:**	
Sl. No		Yes/No/NA	Remarks
1	Ensure that all pre-commissioning checks has been made		
2	Pressure testing of pipework was carried out before insulation		
3	Pressure testing test reports are available		
4	Verify that each valve has high enough pressure to operate, e.g. pressure independent control and balancing (PICB) valve has at least 35 kPa (5 psi) but less than 350 kPa (SO psi) pressure difference across the valve.		
5	Verify that supply w ater temperature is at design temperature.		
6	Check that all manual shut-off valves are open		
7	Check that all bypass valves are closed		
8	Carry out the balancing of water flow rates where the manual balancing valves are used		
9	In case the system has both the PICB and manual balancing valves, all manual balancing valves must be set to their operating point before PICB valve operation are verified		
10	There is no unusual noise/vibration when valves are wither closed or open		
11	In case any chemicals have beed added into the water, the corret chemical has been used and treatment has been made by authorized company		
12	Pressure and water flow rate measurements needs to be recorded in every control valve and results are attached to the commissioning documents		

Checked by:

Date:

Pipework Pre-Commissioning Check List			
Project:		**Contractor:**	
Sl. No		Yes/No/NA	Remarks
1	Was the 'Authorized Person' in charge who is trained to receive the delivery and is aware of the layout of customers facilities and the customers off load procedure?		
2	The types, sizes and dimensions of the delivered items has been verified once arriving to the site		
3	All pipe lines have been made from correct materials (galvanized Iron GI/ copper/plastic (PEC, PERT)/		
4	Confirm that pipework runs are correctly located and have adequate clearance from other services		
5	In case slope is specified, it is executed correctly		
6	Seismic anchoring installed, if specified		
7	Pipes and valves are labelled/tagged and each pipeline can be adequately identified		
8	All labels/tags are readable and do not need replacing		
9	As built drawings are submitted		
10	All welded joints has been checked before insulation		
11	Correct refrigerant has been used and refrigerant is fully charged		
12	All pipes are sufficiently supported		
13	Pipework and pump are supported separately		
14	There are sufficient amount of by-pass valves to conduct a proper flushing		
15	Pipeline flushing (min 3 t imes) has been done to remove any debris build-up or corrosion		
16	Water samples were taken from last flushing water, they have been tested and results have been attached into commissioning documents		

17	Pressure testing of pipework was carried out before insulation		
18	Isolation, balancing, control and other valves are installed as required		
19	Valves installed in proper direction		
20	Operation of valves is verified and motorized valves are wired and ready to operate		
21	Pressure relief valves that require a positive shut-off are verified to not be leaking when closed at normal operating pressure		
22	Thermometers and gauges are installed as required		
23	Test ports (P/T) installed near all control sensors and as per specification		
24	Air vents (manual or automatic) are installed as specified		
25	Strainers are in place and clean		
26	Flexible connections are installed as specified		
27	Vibration insulators are installed as pre specification		
28	All fill in pipe inlets are fitted with a shut-off valve and lockable cap		
29	Insulation thickness of insulation are as per design documents		
30	Technical data of insulation material is available and fulfils the specification in terms of thermal conductivity, water vaoour resistance, surface emissivitv and insulation orotection from ambient conditions like solar,		
31	Insulation is airtight to avoid condensation and rusting		
32	Insulation material is sufficiently protected & no damages in the material surface		
33	Fire rated material has been used in pipe penetrations where specified and installation is fire/smoke tight		
34	All drainage pipes are properly installed and tested that water flows freely out		
35	Flow switches installed as required		
36	Flow meters installed as required		
37	Flow directions labelled on piping insulation		
38	Expansion tanks verified not waterlogged and system is full of water		
39	Chemical treatment pot installed in proper direction		
40	System filled with clean (as specified) water and treatment chemicals (if specified)		
41	Ensure that the system is pressurized and the make-up water system is operational.		
42	No water or refrigerant leakage detected		

Checked by:

Date:

Ductwork Balancing/Ratio Calculations and Final Measurement							
Project:			**Pump name:**				
Contractor:			**Pump ID #:**				
Zone	Item ID	**Before Balancing**			**After Balancing**		
		Air flow rate (dm3/s)		Measured/ Designed Ratio	Water flow rate (kg/s)		Measured/ Designed Ratio
		Designed	Measured		Designed	Measured	

Checked by:

Date:

Precission Air Conditioning (PAC) Delivery Check List			
Project:		PAC name:	
Contractor:		PAC ID#:	
Sl. No		Value	Remarks if not as per specification
1	Manufacture		
2	Unit dimesions		
3	Site assambled or packaged		
4	DX (evaporator) coil dimensions		
5	Slope towards drain in the stainless steel drain pan		
6	Rated cooling capacity		
7	Orientation of DX (evaporator) coil connections		
8	Refrigerant type		
9	Condender model no.		
10	Condenser type (water or air cooled)		
11	Compressor motor (power, type, volts, etc.)		
12	Cooling tower type (open or closed circuit)		
13	Locked Rotor Amps(LRA): current under starting conditions		
14	Rated Loads Amps(RLA): maximum current under any operating condition		
15	Manufacturer's effiency rating (kW/ton) in standard conditions and IEER		
16	Supply Fan model and type (EC, AC, backward curver, etc)		
17	Air flow rate		
18	Fan motor power		
19	Number of fans per unit		
20	Fan belt type, size and number		
21	Vibration eliminators		
22	Orientation offan section doors		
23	Filter types		
24	Filter sizes and quantity		
25	Orientation of filter section doors		
26	All factory-build measurement instruments in place		
27	All factory-build electrical connections done		
28	Volume control damper for supply/return/fresh air duct conections		
29	Canvas connection in the fan outlet		
30	Test & Warranty certificates for motor/coil/fan		
31	Was the 'Authorized Person' in charge of receiving goods who is trained to receive the delivery and Is aware of the layout of customers facilities and the customers off load procedure?		
32	Unit checked for interior and exterior damages		

Checked by:

Date:

	Precission Air Conditiong (PAC) Operation Data Collection Sheet				
Project:			**PAC name:**		
Contractor:			**PAC ID#:**		
Sl. No		Unit	Designed	Measured	Difference
1	Supply air flow rate				
2	Pressure generation of fan with maximum air flow rate				
3	Static pressure across all filters				
4	Static pressure across cooling coil				
5	VFD/fan motor frequency with maximum air flow rate				
6	VFD/fan motor frequency with minimum air flow rate				
7	Fan RPM with maximum air flow rate				
8	Fan motor RPM with maximum airflow rate				
9	Fan motor voltage with maximum air flow rate				
10	Fan motor current with maximum air flow rate				
11	Center-to-center distance of the motor base travel				
12	Compressor motor voltage, phase-to-phase				
13	Voltage imbalance				
14	Compressor motor current (Rated Load Amps RLA)				
15	Main isolator current				
16	Noise level of PAC				
17	Supply air temperature (cooling coil valve fully open)				
18	Return air temperature (befor e cooling coil)				
19	Air temperature in the raised floor				
20	Average (48 h) room air temperature in cold island				
21	Average (48 h) room air temperature in hot island				
22	Ambient air temperature at condenser coil				
23	Supply air relative humidity (cooling coil valve fully open)				
24	Return air relative humi dity (before cooling coil)				
25	Relative humidity in the raised floor				
26	Average (48 h) room air relative humidity in cold island				
27	Average (48 h) room air realtive humidity in hot island				

Checked by:

Date:

	Precission Air Conditioning (PAC) Pre-commissioning Check List		
Project:		**PAC name:**	
Contractor:		**PAC ID#:**	
Sl. No		Yes/No/NA	Remarks
1	Correct PAC is installed and located as per drawing		
2	Clear access to all required components		
3	All transit blocks are removed		
4	Supply & return ducts are connected and complete		
5	Canvas between unit and ductwork is installed correctly		
6	Ensure that air flow and static pressure measurement sensors are located in a straight section of duct (not right after damper, branch or damper)		
7	All pipe work connections are completed to all coils		
8	All cooling coils are secured and f ins combed		
9	Pipe supports are sufficient		
10	All capillary t ubes are tied down to prevent excess vibration		
11	All valves, gauges, thermostat and monitoring instruments are installed		
12	Proper condensate drain piping is installed with slopes and supports		
13	Refrigerant pipe insulation work is done as per design drawings		
14	All filters are as per design and installed correctly		
15	Temporary filter media is installed covering air intake and return connections to protect main filters		
16	Fans and motors rotates properly to the right direction		
17	Fan belts are fitted correctly		
18	Fan shaft and bearings are aligned		
19	Pulleys are aligned and belt tension is correct		
20	Cleanliness of the bearings, lubricant is fresh and of the correct grade		
21	Coolant is available at the bearings (if specified)		
22	Drive guards are fitted and access for speed measurement is provided		
23	Ensure satisfactory operation of the inlet guide vanes over full range of movement		
24	Check that fan casings is earthed correctly and soundly bonded.		
25	Fused disconnect switch sized in accordance with the requirements		
26	Proper size fuses in the disconnected switch		
27	Any anti vibration mountings required are installed		
28	All electrical isolators are in place and connections complete and tested to motors		
29	Electrical cabling connections are with cable tray		
30	All controls are installed, calibrated and fully operational		
31	Variable Frequency Driver {VFD) is installed and operational		
32	Proper size fuses are used and fuse disconnect switch is sized in acording to the requirements		
33	Blanking of gaps are done where necessary with sealant		
34	Lighting within units is in place and connected		
35	Dampers are free moving through full range		
36	Appropriate safety warning signs are in place		
37	PAC is free from damages/scratches		
38	All components, ducts and pipes as well as AHU rooms are clean		
39	There is no oil or refrigerant leak in the PAC unit		
40	Raised floor is clean and air tight		
41	Supply air diffusers in the raised floor are in correct type, size and location		
42	Return air system is as per drawings		
43	Power is available during commissioning		
44	BMS data is available during the commissioning		
45	PAC system is in operation before the test and balance procedures are started		
46	PAC manufacturer has pressure tested the unit after assambly on site or in case of factory assambled packaged unit, the pressure testing certificate has been handed over to the contractor		

Checked by:

Date:

Precission Air Conditioning (PAC) Start-up Check List			
Project:		**PAC name:**	
Contractor:		**PAC ID#:**	
SI. No		Yes/No/NA	Remarks
1	PAC pre-commissioning checks are completed		
2	Electrical tests are complete & available from electrical contractor		
3	Ensure that the ductwork is properly balanced before making operational measurements in PAC and provide air flow measurment tables and ratio calculations in case supply side is ducted		
4	Ensure that static pressure control of fan operates properly		
5	Check the fan for proper operating condition and ensure that the fan motor is below the full-load current		
6	Determine fan airflow by measuring the air velocity in front of the filter section. Air flow rate should be 100? 110% of designed air flow rate - compare the air flow rate also with the results on the main ducts after the AHU to ensure there is no air leakage		
7	Measure the air flow rates in all supply air diffusers in the floor and confirm that the discrepancy is within		
8	Measure static pressure difference across the fan section (pressure creation) and across all other components in the PAC (pressure loss)		
9	Define the operation point of fan in the fan curve and attach it into the commissioning documents		
10	Check the fan and motor frequency, voltage, currency and RPM		
11	Measure the cooling systems operation parameters		
12	Measure the room air temperature and RH during 48 h inside the data center both in the cold and hot isles		
13	Measure outdoor, supply and return air quality parameters		

Checked by:

Date:

Split, Multi-split and VRF AC Delivery Check List			
Project:		**Model ID#:**	
Contractor:			
SI. No		Value	Remarks if not as per specification
1	Manufacture		
2	Types of AC system de livered is same as per requirement (capacity, model, ec)		
3	The packing has been properly made and there are no damages to the product		
4	The sub·accesorrie s should also be with the product (filters, installation, hanging plate)		
5	Rated cooling capacity		
6	Rated heating capacity		
7	All wiring and piping connections have been made at factory		
8	Installation manual is included in the box		
9	Test & Warranty certificates		
10	Pipe connections are correct in size		
11	Was the 'Authorized Person' in charge of receiving goods has been trained to receive the delivery and is aware of the layout of customers facilities and the customers off load procedure?		
12	AC units received as ordered		

Checked by:

Date:

SI. No	Split, Multi-Split and VRF AC Pre-commissioning Check List	Yes/No/NA	Remarks
Project:		Model ID#:	
Contractor:			
SI. No		Yes/No/NA	Remarks
1	Delivery checks have been made		
2	All connections (pipe, condensation hose, electrical) has been correctly made		
3	Refrigerated pipe work has been pressure tested		
4	Ensure that the system has been properly evacuated, dehydrated and charged		
5	Correct refrigerant trim charge has been calculat ed and added		
6	Check that the units are correctly wired up		
7	Ensure that the connect ion nuts are tightened		
8	All units, remote controllers and centralised controllers have been correctly addressed prior to turning on the power to the outdoor unit		
9	All condensat ion drain pipe work must be completed and tested		
10	Outdoor unit is correctly installed		
11	In case of heat pump, the outdoor unit condensat ion (heating mode) water piping is provided		
12	Ensure that air filters are properly installed		
13	Appropriate safety warning signs in place		
14	Confirm power availability for duration of commissioning		
15	BMS data is available during the commissioning		
16	AC system is in operation before the test procedures are started		

Checked by:

Date:

SI. No	Split, Multi-Split and VRF AC Pre-commissioning Check List	Yes/No/NA	Remarks
Project:		Model ID#:	
Contractor:			
SI. No		Yes/No/NA	Remarks
1	Delivery checks have been made		
2	All connections (pipe, condensation hose, electrical) has been correctly made		
3	Refrigerated pipe work has been pressure tested		
4	Ensure that the system has been properly evacuated, dehydrated and charged		
5	Correct refrigerant trim charge has been calculated and added		
6	Check that the units are correctly wired up		
7	Ensure that the connection nuts are tightened		
8	All units, remote controllers and centralised controllers have been correctly addressed prior to turning on the power to the outdoor unit		
9	All condensation drain pipe work must be completed and tested		
10	Outdoor unit is correctly installed		
11	In case of heat pump, the outdoor unit condensation (heating mode) water piping is provided		
12	Ensure that air filters are properly installed		
13	Appropriate safety warning signs in place		
14	Confirm power availability for duration of commissioning		
15	BMS data is available during the commissioning		
16	AC system is in operation before the test procedures are started		

Checked by:

Date:

	Split, Multi-Split and VRF AC Operation Data Collect ion Sheet					
Project:				**Model ID#:**		
Contractor:						
SI. No		Unit	Designed	Measured	Difference	
1	Room air temperature					
2	Supply air temperature					
3	Voltage					
4	Starting current					
5	Running current with maximum speed					
6	High pressure cut out					
7	lowpressure cut out					
8	Compressor suction pressure					
9	Compressor discharge pressure					
10	Evaporat or entering DB temperature					
11	Evaporat or entering WB temperature					
12	Evaporator leaving DB temperature					
13	Evaporator leaving WB temperature					
14	Condenser entering OB temperature					
15	Condenser entering WB temperature					
16	Condenser leaving DB temperature					
17	Condenser leaving WB temperature					
18	Air flow rate					
19	Noise level					
Checked by:						
Date:						

Checklists for Client and/or Commissioning Provider

	AHU/TFA Inst allation Review Check List		
Project:		**Cx Provider:**	
SI. No		Yes/No/NA	Remarks
1	Delivery check of all AHU/TFA units has been made and documented		
2	All AHU/TFA units are delivered and installed as per specification		
3	Units were protected from dirt and water during the storage on site		
4	Units are clean and no there are no damages inside or outside the units		
5	All TFA/AHU units has been pressure tested either in the factory (packaged) or on site (site assambled) and test sertiflcates are available		
6	Units and all access doors are air tight and joints are sufficiently sealed and all selants are not damaged		
7	Filters are as per specification and installed correctly		
8	All water coils and pipes are properly installed, insulated and pipes are sufficient ly supported		
9	Condensation tray and pipes are properly installed		
10	All measurement and monitoring instruments are on correct place and operational		
11	Fan is correctly installed and electrical connections are properly made		
12	Vibration isolators are on place		
Checked by:			
Date:			

AHU/TFA Start Up and Operation Review Check List			
Project:		Cx Provider:	
Sl. No		Yes/No/NA	Remarks
1	Pre-commissioning checks has been made for all AHU/TFA units		
2	Contractor has checked that all start-up activities have been made		
3	Ductwork balancing was made before operational measurements and balancing records are available		
4	All air flow and pressure measurements are carried out		
5	Actual fan operation point is marked into the fan curve		
6	All water side measurements has been carried out		
7	Cx provider was whitnessIng the AHU/TFA measurements		
8	Cx provider has reviewed all AHU/TFA measurement results		
9	Energy recovery wheel measurements has been done and the efficiency has been calculated		
10	Outdoor and supply air quality measurements has been carried out and supply air quality is as per specification		
11	There is no unusual noise or vibration		
12	Correct and calibrated instruments have been used during performance measurements		
13	AJI AHU/TFA units operate as per specification		
Checked by:			
Date:			

Exhaust Fan Installation Review Check List			
Project:		Cx Provider:	
Sl. No	Description	Yes/No/NA	Remarks
1	Delivery checks of all exhaust fans have been made and documented		
2	All fans are delivered and installed as per specification		
3	Fans were protected from dirt and wate' durine the storage on site		
4	Fans are clean and no there are no damages Inside or outside the unit		
5	Fans are correctly installed and electrical connections are properly made		
6	Vibration isolators are on place		
Checked by:			
Date:			

Exhaust Fan Start up and Operation Review Cheek List			
Project:		Cx Provider:	
Sl. No	Description	Yes/No/NA	Remarks
1	Pro-commissioning checks has been mad for all fans		
2	Contractor has checked that all start-up activities have been made		
3	All air flow and pressure measurements are carried out		
4	Actual fan opertation point is marked into the fan curve		
5	Cx provider was whitnessing the fan measurements		
6	Cx provider has reviewed all ten measurement results		
7	There are no unusual noise or vibration		
8	Correct and calibrated instruments have been used during performance measurements		
9	All exhaust fans operate as per specification		
Checked by:			
Date:			

VFD Installation Review Check list

SI. No	Description	Yes/No/NA	Remarks
Project:		Cx Provider:	
SI. No	Description	Yes/No/NA	Remarks
1	Delivery checks of all VFDs have been made and documented		
2	All VFDs are delivered and installed as per specification		
3	VFD were protected from dirs and water during the storage on site		
4	VFDs are correctly installed and electrical connections are properly made		

Checked by:

Date:

VFD Start Up and Operation Review Check List

SI. No	Description	Yes/No/NA	Remarks
Project:		Cx Provider:	
SI. No	Description	Yes/No/NA	Remarks
1	Pre-commissioning checks has been made for all VFDs		
2	Contractor has checks there all start-up activities have been made		
3	All performance measurements are carried out		
4	Cx provide has reviewed all operational measurement results		
5	There are no unusual noise		
6	All VFDs operate as per specification		

Checked by:

Date:

Ductwork Installation Review Check List

SI. No	Description	Yes/No/NA	Remarks
Project:		Cx Provider:	
SI. No	Description	Yes/No/NA	Remarks
1	Precommissioning checks are completed		
2	Dues are made from correct material as specified in design documents		
3	All bents, T-branches, reductions and expansions are made so that they do not increase unnecessarily pressure loss or create the additional leakage		
4	All joints are sequered with aluminium duct tape		
5	Ducts are sufficiently supported (type, distance, etc.)		
6	Ductwork leakage test (as specified in design documents) was completed before insulation		
7	Cx provider witnessed the ductwork leakage testing		
8	Ductwork leakage tests were all 'PAST'		
9	Ensure that all ducts are clean inside All Volume Control Dampers (VCD), VAV dampers and static pressure dampers are correctly installed		
10	Fire dampers, smoke dampers, and access doors are installed as designed		
11	Sound attenuators are properly made and do not release Man-made Vitorous Fibres IMMVF) into the air		
12	Technical data of insulation material is available and fulfills the specification in terms of thermal conductivity, water vapour resistance, surface emissivity and insulation protection from ambient conditions like solar, temperature and rain		
13	Ensure that insulation is properly made (fully air tight, no damages in the surface, no air pockets between insulation and duct, etc.) and protected from weather in outdoor installation		
14	As-built drawings has been submitted		

Checked by:

Date:

Ductwork Start-up and Operation Review Check List

Project:			Cx Provider:	
Sl. No		Description	Yes/No/NA	Remarks
1		Ensure that all pre-commissioning and start-up procedures has been made		
2		Leakage testing of ductwork was carried out before insulation		
3		The ductwork leakage test reports are available		
4		Ensure that all ducts are clean inside		
5		Ensure that ductwork is properly balanced		
6		Cx Provider was whitnessing the ductwork balancing work		
7		Cx Provider has reviewed all ductwork balancing records (ratio calculations, AHU before & after measurement sand final air flow measurements)		
8		Ensure that there is not too much noise generated by the ventilation system		
9		Review the air flow rate measurement of each VAV damper with maximum and minimum position		
10		Correct and calibrated instruments have been used during performance measurements		
11		All ductwork components operate as per specification and correct air flow rates are achieved in each space without unnecessary drought in the occupied zone		

Checked by:

Date:

Staircase Pressurization Fan Installation Review Check List

Project:			Cx Provider:	
Sl. No			Yes/No/NA	Remarks
1		Delivery checks of all staircase pressurization fans have been made and documented		
2		All fans are delivered and installed as per specification		
3		Fans were protected from dirt and water during the storage on site		
4		Fans are clean and no there are no damages inside or outside the unit		
5		Fans are correctly installed and electrical connections are properly made		
6		Vibration isolators are on place		

Checked by:

Date:

Staircase Pressurization Fan Start Up and Operation Review Check List

Project:			Cx Provider:	
Sl. No			Yes/No/NA	Remarks
1		Pre-commissioning checks has been made for all fans		
2		Contractor has checked that all start-up activities have been made		
3		All air flow and other performance measurements of each fan are carried out		
4		Cx provider was wit nessing the fan measurements		
5		Pressure differences across doors and forces to open the doors have been measured		
6		Cx provider was witnessing the pressure difference and opening force measurements		
7		There are no unusual noise or vibration		
8		Correct and calibrated instruments have been used during performance measurements		
9		All staircase pressurization fans operate as per specification		

Checked by:

Date:

Chiller Installation Review Check List			
Project:		**Cx Provider:**	
Sl. No		Yes/No/NA	Remarks
1	Delivery check of all chillers have been made and documented		
2	All chillers are delivared and installed as per specification		
3	Units were protected from dirt and water during the storage on site		
4	Units are clean and no there are no damages inside or outside the units		
5	All pipe connection have been completed		
6	All measurement and monitoring instruments are no correct place and operational		
7	Correct infrigerant has been used		
8	Vibration isolators are on place		
Checked by:			
Date:			

Chiller Start Up and Operation Review Check List			
Project:		**Cx Provider:**	
Sl. No		Yes/No/NA	Remarks
1	Pre-commissioning checks has been made for all chillers		
2	Contractor has checks there all start-up activities have been made		
3	All performance measurements are carried out		
4	Cx provide was witnessing chiller measurements		
5	Cx provide has reviewed all chillers measurement results		
6	There are no unusual noise		
7	Correct and calibrated instruments have been used during performance measurements		
8	All chillers operate as per specification		
Checked by:			
Date:			

Pump Installation Review Check List			
Project:		**Cx Provider:**	
Sl. No	Description	Yes/No/NA	Remarks
1	Delivery checks of all pumps have been made and documented		
2	All pumps are delivered and installed as per specification		
3	Pumps were protected from dirt and water during the storage on site		
4	Pumps are clean and no there are no damages inside or outside the unit		
5	Pumps are correctly installed and electrical connections are properly made		
6	Pumps are properly installed if specified		
7	Vibration isolators are on place		
Checked by:			
Date:			

Pump Start Up and Operation Review Check List			
Project:		**Cx Provider:**	
Sl. No	Description	Yes/No/NA	Remarks
1	Pre-commissioning checks has been made for all pumps		
2	Contractor has checked that all start-up activities have been made		
3	All water flow and pressure measurements are carried out		
4	Actual pump operator paint is marked in the pump curve		
5	Cx provider was whitnessing the pump measurements		
6	Cx provider was whitnessing all pump operation measurement results		
7	There are no unusual noise or vibration		
8	Correct and calibrated instruments have been used during performance measurements		
9	All pumps operate as per specification		

Checked by:

Date:

Cooling Tower Installation Review Check List			
Project:		**Cx Provider:**	
Sl. No		Yes/No/NA	Remarks
1	Delivery checks of all cooling towers have been made and documented		
2	All cooling towers are delivered and installed as per specification		
3	Cooling towers were protected from dirt and water during the storage on site		
4	Cooling towers are clean and no there are no damages inside or outside the unit		
5	All components that are specified have been installed		
6	Cooling towers are correctly installed and electrical connections are properly made		
7	Vibration isolators are on place		

Checked by:

Date:

Cooling Tower Start Up and Operation Review Check List			
Project:		**Cx Provider:**	
Sl. No		Yes/No/NA	Remarks
1	Pre-commissionine checks has been made for all cooline towers		
2	Cont ractor has checked that all start-up activities are completed		
3	All performance measurements are carried out		
4	Cx provider has reviewed the performance results		
5	Water quality has been tested and is acceptable		
6	There are no unusual noise or vibration		
7	Correct and calibrated instruments have been used during performance measurements		
8	All cooling towers operate as per specification		

Checked by:

Date:

Pipework Installation Review Check List			
Project:		**Cx Provider:**	
Sl. No		Yes/No/NA	Remarks
1	Precommissioning checks are completed		
2	Pipes are made from correct material as specified in design documents		
3	Pipeworks run in correct locations and each one of them can be identified from labels/tags		
4	All joints are water tight		
5	Pipes are sufficiently supported (type, distance, etc.)		
6	Pipes were sufficiently flushed, water sample of last flushing water was clean and test results are available		
7	Strainers are cleaned after flushing		
8	Pipework pressure testing was completed before insulation		
9	CxA witnessed the pipework pressure testing		
10	Pipework pressure tests were all 'PAST'		
11	Correct refrigerant has been used and refrigerant is fully charged		
12	Technical data of insulation material is available and fulfills the specification in terms of thermal conductivity, water vapour resistance, surface emissivity and insulation protection from ambient conditions like solar, temperature and rain		
13	Ensure that insulation is properly made (fully air tight, no damages in the surface, no air pockets between insulation and duct, etc.) and protected from weather in outdoor installation		
14	All valves has been correctly installed (location, safety distance, direction, access)		
15	No water or refrigerant leakage was detected		
16	As-built drawings has been submitted		
Checked by:			
Date:			

Pipework Start-up and Operation Review Check List			
Project:		**Cx Provider:**	
Sl. No		Yes/No/NA	Remarks
1	Ensure that all pre-commissioning and start-up procedures has been made		
2	Pressure testing of pipework was carried out before insulation		
3	The ductwork thightness test reports are available		
4	Ensure that pipework is properly balanced		
5	Cx provider was witnessing the pipework balancing work		
6	Cx provider has reviewed all pipework balancing records (ratio calculations, water flows before & after)		
7	Review the water flow rate measureme nt of each valve has been made with a ll valves fully open as well as in the situation where only e.g. 10% of valves are open		
8	Ensure that there is not too much noise generated by the hydraulic system		
9	In case any chemicals have beed added into the water, the corret chemical has been used and treatment has been made by authorized company		
10	Correct and calibrated instruments have been used during performance measurements		
11	All pipework components o perate as per specification and correct water flow rates are achieved.		
Checked by:			
Date:			

\multicolumn{4}{c}{**Precission Air Conditiong (PAC) Installation Review Check List**}			
Project:		**Cx Provider:**	
SI. No		Yes/No/NA	Remarks
1	Delivery check of all PAC units has been made and documented		
2	All PAC units are delivered and installed as per specification		
3	Units were protected from dirt and water during the storage on site		
4	Units are clean and no there are no damages inside or outside the units		
5	All PAC units has been pressure tested either in the factory (packaged) or on site (site assambled) and test sertificates are available		
6	Units and all access doors are air tight and joints are sufficiently sealed and all selants are not damaged during installation		
7	Filters are as per specification and installed correctly		
8	All pipes are properly installed, insulated and pipes are sufficiently supported		
9	Condensation tray and pipes are properly installed		
10	All measurement and monitoring instruments are on correct place and operational		
11	Fan is correctly installed and electrical connections are properly made		
12	Vibration isolators are on place		
13	Raised floor is clean and air tight		
14	Supply air diffusers in the raised floor are in correct type, size and location		
15	Return air system is as per drawings		

Checked by:

Date:

\multicolumn{4}{c}{**Precission Air Conditioning (PAC) Start Up and Operation Review Check List**}			
Project:		**Cx Provider:**	
SI. No		Yes/No/NA	Remarks
1	Pre-commissioning checks has been made for all PAC units		
2	Contractor has checked that all start-up activities have been made		
3	Ductwork balancing was made before operational measurements and balancing records are available		
4	All air flow and pressure measurement s are carried out		
5	Actual fan operation point is marked into the fan curve		
6	All water side measurements has been carried out		
7	Cx provider was whitnessing the PAC measurements		
8	Cx provider has reviewed all performance measurements		
9	All performance measurements have been carried out and they are as per specification		
10	All electrical measurements have been carried out		
11	There is no unusual noise or vibration		
12	Correct and calibrated instruments have been used during performance measurements		
13	All PAC unit s operate as per specif ication		

Checked by:

Date:

Split, Multi-Split and VRF AC Installation Review Check List			
Project:		**Cx Provider:**	
Sl. No		Yes/No/NA	Remarks
1	Delivery check of a ll AC units have been made and documented		
2	All AC units are delivered and installed as per specification		
3	Units were protected from dirt and water during the storage on site		
4	Units are clean and no there are no damages inside or outside the units		
5	All pipe connections have been completed		
6	Correct refrigerant has been used		
Checked by: Date:			

Split, Multi-Split and VRF AC Start Up and Operation Review Check List			
Project:		**Cx Provider:**	
Sl. No		Yes/No/NA	Remarks
1	Pre-commissioning checks has been made for all AC units		
2	Contractor has checked that all start-up activities have been made		
3	All performance measurements are carried out		
4	Cx provider was witnessing performance measurements		
5	Cx provider has reviewed all performance measurement results		
6	There is no unusual noise or vibration		
7	Correct and calibrated instruments have been used during performance measurements		
8	All AC units operate as per specification		
Checked by: Date:			

REFERENCES

1. Air Conditioning Contractors America (ACCA), 2010, HVAC Quality Installation Specification, US.

2. ANSI/ASHRAE Standard 55, (2013), Thermal Environmental Conditions for Human Occupancy.

3. ANSI/ASHRAE 52.2, (2012), Method of Testing General Ventilation Air-Cleaning Devices for Removal Efficiency by Particle Size.

4. ANSI/ASHRAE Standard 62.1 (2013), Ventilation for Acceptable Indoor Air Quality.

5. ANSI/ASHRAE/IES Standard 202-2013, Commissioning Process for Buildings and Systems.

6. ASHRAE, (2013), Guideline 0, The Commissioning Process.

7. ASHRAE, (2015), Guideline 0.2, Commissioning Process for Existing Systems and Assemblies.

8. ASHRAE, (2007), Guideline 1.1, HVAC&R Technical Requirements for The Commissioning Process.

9. ASHRAE, (2014), Guideline 1.4, Procedures for Preparing Facility Systems Manuals.

10. ASHRAE (2012), Guideline 1.5, The Commissioning Process for Smoke Control Systems.

11. ASHRAE (2013), Guideline 4, Preparation of Operating and Maintenance Documentation for Building Systems.

12. ASHRAE (2009), Guideline 11, Field Testing of HVAC Controls Components.

13. ASHRAE (2013), Refrigeration Commissioning Guide for Commercial and Industrial Systems

14. ASHRAE, AICARR, CIBSE (2017), Commissioning Definitions and Terminology for the Building Industry-a common overview.

15. Appraisal Institute, (2013), Green Building and Property Value A Primer for Building Owners and Developers, US.

16. The Australian Institute Of Refrigeration, Air Conditioning And Heating, (2012), Methods of calculating Total Equivalent Warming Impact (TEWI), Australia.

17. Babiak, J. et.al., (2006), Low Temperature Heating and High Temperature Cooling, REHVA Guidebook no. 7, Brussels.

18. The Better Building Partnership, (2011), Better Metering Toolkit, UK.

19. BSRIA, (2001), Pre-commissioning Cleaning of Pipework Systems, BSRIA Guide, UK.

20. BSRIA, (2015), Soft Landing Guidance, UK.

21. Bureau of Energy Efficiency BEE, (2016), https://beeindia.gov.in/, Ministry of Power, India.

22. Californian Commissioning Collaborative, (2006), California Commissioning Guide, Existing Buildings, US.

23. Californian Commissioning Collaborative, (2010), California Commissioning Guide, New Buildings, US.

24. CPCB, (2015) Environmental Data Bank, Central Pollution Control Board, Government of India.

25. Cullick, H., (2010), HVAC Commissioning Checklists, BT Collins High Tech RTS, Sacramento, CA, US.

26. Danfoss, (2016), http://www.danfoss.com, Denmark.

27. Design Builder, (2016), http://www.designbuilder.co.uk, UK.

28. EN Standard 779, (2012), Particulate air filters for general ventilation. Determination of the filtration performance.

29. EN Standard 15251, (2007), Indoor Environmental Input Parameters for Design and Assessment of Energy Performance of Buildings Addressing Indoor Air Quality, Thermal Environment, Lighting and Acoustics.

30. Energy Star, (2007), Retro-commissioning, US.

31. Environmental Protection Agency, (2009), EPA Building Commissioning Guidelines, Facilities Management & Services Division Office of Administration & Resources Management, US.

32. General Services Administration GSA, (2004), LEED Cost Study, Final Report, US.

33. General Services Administration GSA, (2005), The Building Commissioning Guide, US.

34. The Government Of The Hong Kong Special Administrative Region, (2007), Testing And Commissioning Procedure For Air-Conditioning, Refrigeration, Ventilation And Central Monitoring & Control System Installation In Government Buildings Of the Hong Kong Special Administrative Region.

35. Gupta, N.C. (2016) Comprehensive HVAC System Design, Viva books.

36. HEPAC-website, (2016), www.hepac.com.

37. IARC and WHO, (2013), Outdoor air pollution a leading environmental cause of cancer deaths, press release no. 221.

38. ISHRAE, (2015), Position Paper on Indoor Environmental Quality, India.

39. ISHRAE, (2016), Position paper on refrigerants for Indian refrigeration and air conditioning industries: challenges and responsibilities, India.

40. ISHRAE Standard (2016), Indoor Environmental Quality.

41. ISHRAE-RAMA, (2015) Guidelines for testing and rating of liquid-chilling packages..

42. ISO standard 16890-1, (2016), Air filters for general ventilation -- Part 1: Technical specifications, requirements and classification system based upon particulate matter efficiency (ePM).

43. Jagemar, L., et.al., (2007), The EPBD and Continuous Commissioning, Intelligent Energy, Europe.

44. Knight, A. & Blatch, S., (2014), Final report documenting the final version's algorithms and methodology, Inspection of HVAC Systems through continuous monitoring and benchmarking Intelligent Energy Europe Project Number: IEE-10-272, iSERV, Brussels.

45. Mills, E.P., (2010), A Golden Opportunity for Reducing Energy Costs and Green House Gas Emissions, Lawrence Berkeley National Laboratory, US.

46. Mills, E.P. & Mathew P.P. (2009), Monitoring-Based Commissioning: Benchmarking Analysis of 24 UC/CSUIOU Projects, Lawrence Berkeley National Laboratory, US.

47. Mills, E., et.al., (2004), The Cost-Effectiveness of Commercial-Buildings Commissioning, Lawrence Berkeley National Laboratory, US.

48. Muller, D. et.al., (2013), Mixing Ventilation – A guidebook on mixing air distribution design, REHVA Guidebook no. 19, Brussels.

49. NADCA ACR, (2006), Assessment, Cleaning, and Restoration of HVAC Systems, An Industry Standard Developed by the National Air Duct Cleaners Association.

50. National Science and Technology Council Committee on Technology, (2011), Sub-metering of Building Energy and Water Usage Analysis and Recommendations of the Subcommittee on Buildings Technology Research and Development, US.

51. National Environmental Balancing Bureau NEBB, (2009), Procedural Standards for Whole Building Systems Commissioning of New Construction, US.

52. NIBS Guideline 3, (2012), Building Enclosure Commissioning Process BECx, US.

53. Nonnenmann, J., Chilled Water Plant Pumping Schemes, US.

54. NSW and Office of Environment and Heritage, (2012), Measurement and Verification Operational Guide Commercial Heating, Ventilation and Cooling Applications, Australia.

55. OECD, (2014), The Cost of Air Pollution, Europe.

56. Orn, R., (2010), Building Commissioning – Advantages and disadvantages of the process and how it has been applied in Denmark, DTU Management, Denmark.

57. Pasanen, P. et.al., (2007), Cleanliness of Ventilation System, REHVA Guidebook no. 8, Brussels.

58. The Portland Energy Conservation, Inc. PECI, (1998), Rebuild America Program, Department of Energy's (DoE), US.

59. Raghavan, K. (2016), Building management system, ISHRAE Journal.

60. Skistad, H., et.al., (2001), Displacement Ventilation in Non-industrial Premises, REHVA Guidebook no. 1, Brussels.

61. SMACNA, (1985), HVAC Duct Construction Standards, Metal and Flexible, Second Edition.

62. SMACNA, (1985), HVAC Air Duct Leakage Test Manual.

63. UNEP, (2014), Air Pollution: World's Worst Environmental Health Risk.

64. UNEP, (2014), Sustainability Metrics: Translation and Impact on Property Investment and Management.

65. UNEP, (2011), An Investors' Perspective on Environmental Metrics for Property.

66. UNEP, (2016), Sustainable Real Estate Investment Implementing the Paris Climate Agreement: An Action Framework.

67. USGBC, (2016), LEED v4, http://www.usgbc.org/v4, US.

68. Virta, M., et.al., (2005), Chilled Beam Application Guidebook, REHVA Guidebook no. 5, Brussels.

69. Virta, M., et.al., (2012), HVAC in Sustainable Office Buildings – A bridge between owners and engineers, REHVA Guidebook no. 16, Brussels.

70. Wargorcki, P., et.al (2005), Indoor Climate and Productivity in Offices, REHVA Guidebook no. 6, Brussels.

71. Wattson, R., (2011), Green Building Market and Impact Report.

72. Wellsh, B., (2009), Ongoing Commissioning (OCx) with BAS and Data Loggers, National Conference on Building Commissioning.

73. Wikipedia, (2016), https://en.wikipedia.org.

74. WHO, (2005), Air Quality Guidelines.

75. WHO, Report on a WHO meeting, (2000), The Right to Healthy Indoor Air.

ABOUT ISHRAE

The Indian Society of Heating, Refrigerating and Air Conditioning Engineers (ISHRAE), was founded in 1981 at New Delhi by a group of eminent HVAC&R professionals. ISHRAE today has over 10,000 HVAC&R professionals as members and additionally there are 8,500 Student-members. ISHRAE operates from over 40 Chapters and sub Chapters spread all over India, with HQ in Delhi. It is led by a team of elected officers, who are members of the Society, working on a voluntary basis, and collectively called the Board of Governors.

ISHRAE Objectives:

- Advancement of the Arts and Sciences of Heating, Ventilation, Air Conditioning and Refrigeration Engineering and Related Services.

- Continuing education of Members and other interested persons in the said sciences through Lectures, Workshops, Product Presentations, Publications and Expositions.

- Rendition of career guidance and financial assistance to students of the said sciences.

- Encouragement of scientific research.

ISHRAE Mission

To promote the goals of the Society for the benefit of the general public. Towards this objective, the Chapters of the Society participate in, and organize, activities to protect the Environment, improve Indoor Air Quality, help Energy Conservation, provide continuing education to the Members and others in the HVAC & related user Industries and offer certification programs, career guidance to students at the local colleges and tertiary institutions.

Activities

As part of its objectives to promote the interests of the HVAC&R Industry, ISHRAE is involved in various activities. ISHRAE reaches out to all its members and seeks their active participation & involvement in all the Events/Programs organized by the society.

Programs

ISHRAE conducts Conferences, Seminars, Exhibitions, Workshops, Panel Discussions and Product Presentations throughout the country with both national and international participants to discuss, promote and display the state of the art technologies, systems, products and services.

Publications

ISHRAE publications strive to help its members & the industry keep up-to date with the technical developments, latest trends, and sunrise technologies. ISHRAE Standards, Fundamental books on various topics, safety guidelines, HVAC&R Handbooks and the extremely popular & informative ISHRAE Journal, are a few such publications.

ACREX INDIA

ISHRAE organizes ACREX INDIA, the largest international exposition in South Asia on the Air-Conditioning, Refrigeration, Ventilation and Building services industry. Held annually, ACREX with nearly 500 exhibitors is considered to be a major opportunity to showcase the latest technologies/innovations, and provide a platform for buyer-seller meet, for technical & commercial personnel in the HVAC&R field.

Education & Training

ISHRAE Institute of Excellence (IIE), the educational arm of the Society, is working towards human resource development in the HVAC&R industry in the country by conducting various courses. One of the most important objectives of ISHRAE is Technical Training, and this is done at various levels.

At the apex of the pyramid we have the ICP (ISHRAE Certified Professional) Certification Courses on Clean rooms AC-Design, AC Service and others. At the next level ISHRAE offer a full time Diploma Course for graduate engineers. In addition, at the Chapter level ISHRAE holds several successful training programs, workshops, short term courses and offers e-learning opportunities. ISHRAE is also working with associates in Skill Development activities.

Student Activities

ISHRAE student chapters in more than 150 engineering colleges encourage students to opt for careers in the HVAC&R industry. Knowledge dissemination is done through seminars, quiz contests like aQuest, plant and site visits.

ISHRAE has launched "ISHRAE Job Junction" nationally, providing a platform for leading employers to recruit candidates who are members from ISHRAE student chapters.

K-12 initiative of ISHRAE is focused on school students' contests, in making them aware of subjects like, energy conservation and environmental concerns through drawing competitions, poster design, quiz and planting of trees. Emphasis on STEM education is stressed to inculcate a scientific fervor & help develop these young children into responsible citizens.

Research

ISHRAE promotes research in the field of HVAC&R technology. It offers financial support to Graduate/Post Graduate students, to carry out innovative work on R & D in Technology, Systems, and Processes. ISHRAE partners with Industry & academia to carry out scientific research associated with the HVAC&R Industry.

Search

Provides a unique platform for B2B and B2C users to share their expertise & requirements in an industry specific search engine. We wish to provide unparalleled user experience HVAC & R and Building Services Industry to increase their reach to all concerned in services & trade. This search engine will help promote the Make in India drive, by providing easy referencing to all stakeholders.

Interaction with Govt. Departments and Associate Societies

ISHRAE works in the National interest with various Govt. Ministries/Departments, e.g. in the development of Standards & drafting of NBC for BIS, working on ECBC with BEE, with Ozone Cell of MoEFCC, on refrigerant gases. ISHRAE is a member & active supporter of National Centre for Cold Chain development (NCCD) Ministry of Agriculture & works closely with NCCD on refrigeration.

ISHRAE is also working in close co-operation with other similar Societies & Organizations, both at national and international level, for the promotion and development of issues like Sustainability, Green Buildings, Energy Efficiency, Environmental Responsibility, Indoor Air Quality, Fire & Safety.

Interaction with Think-tanks & NGOs like NRDC, CEEW, TERI, CSE & UN bodies like UNDP/UNEP is a regular feature. ISHRAE is looked upon as a repository of technical knowledge in the HVAC&R & Building Industry field by peer Organizations & the Govt. of India.

More information: www.ishrae.in

ABOUT REHVA

The Federation of European HVAC Associations (REHVA), now 54 years old, is the pan-European umbrella organization joining more than 100,000 HVAC and building services engineers and experts from 27 European countries. REHVA is dedicated to the improvement of health, comfort and energy efficiency in all buildings and communities. It encourages the development and application of both energy conservation and renewable energy sources. REHVA's mission is to promote energy efficient and healthy technologies for mechanical services of buildings, and to disseminate knowledge among professionals and practitioners in Europe and globally. In these areas, REHVA has a significant impact on national and international research initiatives, policy development and implementation, as well as on the associated educational and training programmes.

REHVA's main activity is to promote the development of economical, energy efficient and healthy technologies for mechanical services of buildings. The Board of Directors supervises this work, while the REHVA Technology and Research Committee oversees the activity of the Task Forces realising guidebook projects. Several Task Forces are currently working on REHVA Guidebooks of various topics: Residential heat recovery ventilation; Displacement ventilation; Air filtration in HVAC systems; Quality management for building commissioning, and others. REHVA is delighted to have launched a new joint Task Force with ISHRAE working on the topic "Indoor environmental quality in school buildings". REHVA is looking forward to working on this subject with ISHRAE experts.

REHVA has by today a series of Guidebooks published in Europe, which are the most important tools to diffuse knowledge on latest developments, and advanced technologies providing practical guidance to practitioners. REHVA has published 23 guidebooks to date that are available on the REHVA website. (www.rehva.eu/publications-and-resources/eshop)

REHVA has continuously evolving cooperation with ISHRAE ever since the first Memorandum of Understanding was signed in 2012. Sharing knowledge is also at the core of the collaboration between ISHRAE and REHVA, which has resulted in several joint seminars during ACREX, exchange of articles and REHVA Journal Special issues released for ACREX, as well as the joint work on certain technical topics. REHVA is proud about this latest outcome of our fruitful cooperation: the ISHRAE–REHVA HVAC Commissioning guidebook.

The REHVA Board would like to express its sincere gratitude to ISHRAE for the cooperation and its support to this guidebook. We thank especially Maija Virta, Ole Teisen for their invaluable work that was crucial in getting this guidebook realized. REHVA equally thanks all the experts from ISHRAE and REHVA for their contribution in publishing the HVAC Commissioning guidebook.

Stefano P. Corgnati

REHVA President